Masonry

MAX ALTH

Masonry

A HomeOwner's Bible

DOUBLEDAY & COMPANY, INC., GARDEN CITY, NEW YORK

1982

Dedicated to

Char
Syme
Misch
Mike
Darcy
Mendle
Arrabella

Library of Congress Cataloging in Publication Data

Alth, Max, 1917–
  Masonry.

  (HomeOwner's bible)
  1. Masonry—Amateurs' manuals. I. Title. II. Series.
  TH5313.A48   693′.1
  ISBN 0-385-15399-6 (pbk.)   AACR2
  Library of Congress Catalog Card Number 80–1799
  COPYRIGHT © 1982 BY MAX ALTH

CONTENTS

INTRODUCTION

Working with concrete and masonry—making steps, walks, driveways, patios, foundations and even buildings—is far simpler than most home owners realize. While a beginner will certainly not duplicate the speed and accuracy of a skilled mason, there is no need for speed when you are working for yourself, and in most instances little accuracy is needed. When it is, you can secure the same accuracy by using more care. As far as strength and permanence are concerned, if you follow the general guidelines presented in this book, your work will be as strong and as long-lasting as the work of any other mason, meaning your constructions will outlive us all by several centuries, at least.

Concrete and masonry work is fairly simple. Mainly it is repetitive, which permits you to learn and gain speed as you work. The work is also pleasant if you do not tax yourself by attempting to do too much at one time, or work too fast or too long, and if you remain indoors during inclement weather. Tool and equipment requirements are modest, and these can be obtained from a local "rent-all" shop, if you wish.

Masonry and concrete work is also highly remunerative. The dollar savings between doing it yourself and hiring others can be several times the base cost of the materials alone. The result of your efforts will be a beautiful, permanent addition to your home and grounds. Even if whatever you have constructed is your first attempt at masonry, the results will be attractive. That is the nature of the material.

CHAPTER ONE
Cement and Mortar

CEMENT

Cement is not a new invention. It has been in use for thousands of years. The earliest cements were mixtures of tar and sand, with the addition of clay sometime later. By Roman times, cement was already a common building material, and it is a testament to its durability that many Roman and pre-Roman structures are still standing.

In those days, cement was made by grinding seashells and other pure or nearly pure lime substances into a powder, and then heating the powder until the particles melted and fused to form clinkers. After cooling, the clinkers were reground into another powder. The cement was then used exactly as it is today, for making mortar and concrete.

While the process may appear to be rather simple, it is interesting to note that the early Egyptians, brilliant builders though they were, did not know how to make cement. The same applies to the great civilizations of the Western World, such as the Mayans, Aztecs and Incas. Yet all of these cultures constructed tremendous and beautiful structures without a drop of mortar to bind the stones.

Modern cement differs from ancient cement primarily in its ingredients. The basic method of manufacture remains the same. Also, modern cement will harden in the presence of water; ancient cement did not. This difference is very important. Assume, for example, that you have just completed a driveway and it begins to rain. Rain will damage fresh concrete made with modern cement only to the extent that the raindrops will deform the surface of the concrete. Made with ancient cement, the fresh concrete would be penetrated by the rainwater and be "melted" by it. After the rain, the modern cement drive could be made smooth again, while with ancient cement the concrete would remain thinned and soupy.

Modern cement consists mainly of ground limestone plus a percentage of clay and other ingredients. It was invented in 1824 by Joseph Aspin, a brick mason of Leeds, England. He named his product Portland cement because its color resembled the stone quarried for building purposes on the island of Portland. Aspin made his first batch of cement in an oven. Modern cement makers use giant rotary kilns hundreds of yards long that generate temperatures of up to 2,800° F and more.

Because cement is too weak to be used alone, a certain amount of sand is always mixed with it for

Fig. 1. A beautiful example of what can be accomplished with mortar and stone. This fence is constructed of concrete blocks held together with mortar. Courtesy National Concrete Masonry Association.

practical use, and this combination is properly called mortar. Mortar can be used to join brick to brick, stone to stone, glass to glass, tile to tile, and all combinations thereof. Concrete is a mixture of cement, sand and gravel, and in all cases with all cements, modern and ancient, water is the solvent or activating agent.

There are five basic types of cement in use today, properly labeled as such, plus a number of variations. For almost every purpose involving a residence or small commercial building, Type 1 does fine. Since there may be an occasion when another type is more appropriate, however, they are all listed below.

TYPES OF CEMENT

Type 1   General purpose.
Type 2   Low-hydration (used for heavy structures).
Type 3   High-early (hardens more quickly than the others).
Type 4   Low-heat (used for most massive structures, dams, etc.).
Type 5   Alkali-resistant (used where high-alkali-content water and soils are present).

Fig. 2. The paper bag at the left holds some 94 pounds of Type 1 cement. The bag at the right marked MASONRY CEMENT *contains cement plus some 25% lime and is used for making mortar. This bag of masonry cement weighs only 69½ pounds. Both bags contain exactly 1 cubic foot of material.*

Air-entraining cement. When air-entraining chemicals have been added to cement, the letter *A* is added after the cement's designation and the bag will read Type 1A, Type 2A, etc. These added chemicals cause the cement and the concrete and mortar which may be made with these cements to foam and retain billions of microscopic air bubbles. The effect of these air bubbles is to reduce the water necessary to the mixture, improve its plasticity, meaning that the mixture is easier to work, and increase its resistance to frost and the destructive effects of calcium chloride and sodium chloride (common salt). For these reasons, it is advisable to use these cements when making driveways and other exposed structures in areas where the winters are severe and the cities ruin our vehicles by throwing lots of salt on the roads.

Air-entraining cement is used exactly as any other type of cement, with the exception that it requires about half a gallon less water per bag of cement, as mentioned. If you need or want air-entraining cement and it is not available in premixed form, you can make your own by adding the necessary chemicals. Your local masonry yard usually stocks them, but you must use a machine to mix your mortar and concrete. Mixing by hand is a waste of time: you simply cannot mix it well enough to get the air-entraining results.

Special colors. Types 1, 2 and 1A can be obtained in an almost pure white. Standard cement is a kind of green in the bag and hardens to a kind of green-gray. Type 1 can also be found in a warm brown color that looks much like western adobe on hardening. For best color results with these cements, you must use white sand.

Buying cement. Cement is sold in paper bags that hold exactly 1 cubic foot and weigh about 94 pounds. Since all the types are made to industry standards, the cement of one company is no better or worse than that of any other company. So let price be your guide.

Mortar cement. This is Type 1 cement to which a little lime has been added. The use of mortar cement saves the mason the bother of purchasing, storing and adding lime to his mortar when he wants to use lime mortar. Generally, the ratio of lime to cement is one to four. Lime mortar is used with concrete blocks and brick.

Lime. Masonry lime is slaked lime, which means it has been supplied all the water it can absorb (unslaked lime generates heat and nearly boils when water is added). Masonry lime is similar to garden lime, but it is ground to a very fine powder. If you have some left over, you can dust it on your lawn if you wish. Masonry lime is also sold in bags containing 1 cubic foot; they look like cement bags but weigh about half as much.

Prepackaged mixes. These contain cement, lime and sand. They are economical only when a very small quantity is needed. Otherwise, although very convenient, these packages of premixed mortar are horribly expensive to use.

Storage. As mentioned earlier, cement and lime are sold in paper bags. These bags are by no measure waterproof or water resistant, so you cannot leave them out in the weather. You also cannot place them on the ground. Once the lime or cement absorbs water, the substances begin to change chemically, and the change cannot be reversed. If you must store your bags a long time, place them near the furnace, above the floor in your home. If you do not need to store them more than a week or two, place them on planks supported by concrete blocks and cover them completely with a sheet of plastic. Large sheets of this type can be purchased at the masonry yard.

If you find that the cement or lime has formed lumps which you cannot easily break into fine powder, discard the lumps. The remaining powder can be used, but it must be ultra fine and not sandlike, which would indicate it has absorbed water and is no longer any good.

SAND

Sand obviously needs no explanation, but it must be perfectly clean and free of salt. Silt and salt interfere with the binding and chemical changes that occur in the cement. Use dirty sand, and the mortar or concrete will be weak. When you buy your sand from a commercial source such as a masonry yard or a lumber yard that stocks a few masonry supplies, you can be certain the sand is clean. When you dig your own, or remove some from a beach, you had better test it.

Cleanliness. Place 2 or more inches of sand in the bottom of a bottle. Fill the bottle with water and shake it a bit. Then let it stand till the silt, if any, settles (Fig. 3). If you can see a layer of silt on top of the sand, you must wash the sand. This is done by placing the sand in a mixing box (a large, flat metal or plastic box), tilting the box and letting clean water run over the sand in a steady flow.

To test for salt, place the sample of sand in a clean bottle. Add sufficient water to cover the sand,

Fig. 3. A simple way to test sand for cleanliness. After having been mixed with water and permitted to stand awhile, if a thin layer of silt collected on its surface (note the thin, dark line), this sand should not be used without washing. To test for salt content, taste the water.

slosh it around and then taste the water. If it tastes salty, you must wash the sand.

Kinds of sand. You may have been told that "sharp" sand is best, meaning sand with sharp edges. This is an old mason's tale. All sand works equally well, as long as it is clean and salt free.

WATER

The water you use must be clean enough to drink. If there is any show of color and you are hesitant to use it (if not to drink it), fill a quart bottle with the water and let it stand overnight. If you can see a solid layer of silt on the bottom of the bottle, don't use the water for masonry.

MORTAR

Most masons use a mortar made by mixing two to three parts sand with one part cement for joining stone, glass and other nonporous materials. With porous building units such as block and brick, a little lime is added to the mixture. Lime does several things: it makes the mortar more plastic, which means it is more easily applied, and it makes the mortar stickier, which helps the mortar remain in place. Lime also reduces the strength of the mortar, but this reduction is unimportant for almost all masonry work.

Proportions. Again, the usual ratio is one part cement to two or three parts sand. A mixture with less than two parts sand is weak, as is a mixture with more than three parts sand. For most work, masons use a one-to-two mixture. When a little less strength is needed, a one-to-three mixture or even a one-to-four mixture is used. When lime is added, it is never added at a rate of more than one quarter part lime to one part cement.

Note that none of these ratios is critical. You can use a pail or a shovel to measure. For small quantities, a shovel or even a trowel is of sufficient accuracy.

Tools and equipment. You will need either a mortar box or mortar pan, as well as a mortar hoe, for mixing. You will need a shovel for moving the mortar out of the pan or box or what have you, as well as for mixing. This should be a straight-edged shovel. A mortar board or mortar pan for storing the mortar next to the job, within easy reach of the mason, should also be obtained, as well as a bucket for measuring. On a small job, you can mix the mortar in a wheelbarrow and roll it alongside the work.

Mixing. In most cases, mortar can be mixed by hand or by machine. But before you rush off to rent even a very small power mixer, be advised that a single mason's helper can easily mix sufficient mortar to keep two or three masons going. Since it is inadvisable to keep mortar on hand for more than two

hours, even less on a hot day, one thing you don't want to do is mix up more than you can use. If you do, you will have to discard it. So your first step is to decide approximately how much mortar you will require per two-hour period. This is easy if you hand-mix; just mix a little at a time and vary each batch when you see how it goes.

Hand mixing is easy if you don't try to mix the mortar in a "tea cup." At the very minimum you should use a mortar pan. Anything smaller, as for example a bucket, makes for a tremendous waste of effort.

For any quantity larger than a cubic foot at a time, use a mortar box or a concrete drive or cellar floor. Start by measuring and spreading the sand you are going to use over the concrete drive or the box bottom in a three- or four-inch-thick layer. Then spread the cement and lime on top and mix thoroughly until the mixture is of one even color. When mixing in a box, begin adding the water at one end of the box and push or pull the mixture into the water. When mixing on a drive, spread the mix to form an open-center circle. Push or pull the mix into the water poured into the center of the circle.

You will need roughly about five or six gallons of water for each bag of cement, regular or mortar, you mix. Do not, however, try to premeasure the water. The quantity of water needed will vary with the moisture content of the sand and your particular

Fig. 4. Mortar can be mixed in any convenient container. For small quantities, a mason's wheelbarrow is fine.

Fig. 5. For mixing large quantities of mortar, a power mixer is used. Note that the mixer has been raised on blocks so that its contents can be dumped into the mortar box for convenience.

mortar needs. Instead, add a little water at a time, preferably from a bucket—you can control the quantity much better this way—and mix only a portion of the mortar at a time. Keep the portion you are mixing on the sloppy side (a little too much water) this makes for easy mixing. Add more dry material as you go. You can mix the entire batch at one time if you wish, or quit at any time, leaving some of the mix dry.

Most important is the water content of the mix, or that portion of the mix you will use. There must be sufficient water for the cement to absorb all it needs to. To be certain of this, let the mix stand for ten minutes or so. If it doesn't stiffen up, you have sufficient water. If it does, you must add a little water. Note that at this point just a little water can be too much, so add very carefully and mix thoroughly. A proper consistency is that of stiff oatmeal: you should be able to form a little pile without the mortar spreading into a blob.

In the summer when the bricks or blocks are very dry, you will want the mix a little wetter. On damp days you will want less water. The correct mixture is something you will soon learn as you work. Just make certain the mortar is thoroughly mixed and that all of it is equally wet.

Keeping the mortar workable. After the mortar is mixed in a mortar box or a mortar pan, it is then brought over to the work and deposited on a mortar board, which is literally a large, flat board. With the passage of less time than you might imagine, the unused mortar will begin to stiffen. When it does, splash a little water on it and, with your trowel, work the water into the mortar. Whatever mortar remains on the board more than two hours should be discarded. Mortar that falls to the ground should

Fig. 6. Mortar that has stiffened on the mortar board, as shown, or in a mortar pan, should be retempered before using. Add a little water and remix. Discard the mortar if it has been sitting two hours or so.

not be used for anything more than fill. Never use it for joints. It may have picked up dirt.

Estimating needed quantity. The quantity of mortar necessary can be estimated from the number of bricks or blocks you plan to lay. This is discussed in Chapters 3 and 4. The quantity of mortar that results when mixing is easily determined: it is never much more than 1 cubic foot of mortar for every cubic foot of sand used. The reason for this seeming loss of cement, lime and water is that the substances fit into the voids or spaces between the particles of sand. Thus, to mix 1 foot of cement with 3 feet of sand is to end up with only 3 feet of mortar.

CHAPTER TWO

Concrete

Concrete is today without question far and away the most used building material of all. It is comparatively inexpensive and is used for everything from roads, dams, bridges, buildings, giant stadium roofs to boat hulls and garden furniture. It is rot, insect and fireproof, and is virtually indestructible. No one really knows how long concrete can withstand the elements: 2,000-year-old structures are still standing.

Despite its popularity and use over the last two millennia, concrete remains to some degree a mysterious substance. Scientists do not know exactly what makes it do what it does. It begins as a dry mixture, turns into a kind of mud when water is added, and then hardens into a kind of stone that grows stronger with time. Month-old concrete is roughly twice as strong as seven-day-old concrete, and for some twenty-seven years following, the strength of the concrete continues to increase at a diminishing rate.

It isn't necessary to know all there is to be known about concrete in order to work with it successfully, but a general understanding of its nature will ease the work and insure success.

Strength. Concrete has tremendous compressive strength. A 1-inch cube of concrete can support some 3,000 to 5,000 pounds without fracturing (Fig. 7). Its tensile strength (resistance to bending), however, is very low. Typically, concrete has a rated tensile strength of only 500 psi (pounds per square inch). If you were to make a 1-inch-square rod out of concrete, you could break it over your knee. All this simply means that you do not have to even consider how much weight you pile on top of concrete, but when you want the concrete to span an open space, as for example when making a porch, you must place steel bars within the concrete to keep the concrete from cracking and breaking.

The actual strength of concrete depends on the mix—the ratio of cement to sand to stone, and the quantity of water used: the high sand and cement

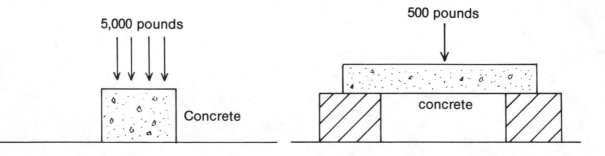

Fig. 7. *Concrete's compressive strength* (*resistance to pressure*) *is easily ten times greater than its tensile strength* (*resistance to bending*).

(6)

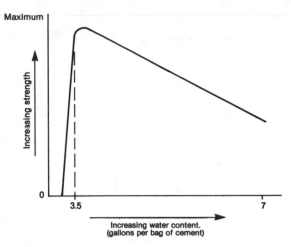

Fig. 8. *As this graph shows, the strength of concrete decreases as you increase the quantity of water in the mix. If insufficient water is used, however, the concrete has almost no strength at all. Thus, it is always better to use a little more water than a little less. Five to six gallons of water to each bag of cement is the ratio commonly used.*

mixes produce the strongest concrete (sand and cement alone, however, are never as strong as when stone is included).

Further, the less water added to a given mixture, the stronger the final concrete, although this principle only goes so far (Fig. 8). Add insufficient water and you don't have weak concrete, you have no concrete. The cement remains dry in parts of the mix, and some of it doesn't receive sufficient water to complete its chemical transformation. Such concrete can be broken by hand.

Maximum concrete strength is secured by using about 3½ gallons of water to every cubic foot (one bag) of cement in the mix, but this is not really practical. Doubling the water only reduces the final strength by one third, and it insures that the mix will receive all the water necessary for proper lubrication, making it far easier to handle, place and level. Most important, the resultant "wet" mix will still be several times stronger than required for any structure you may wish to build, including buildings ten stories high. Thus, a ratio of 5 to 6 gallons of water per bag of cement is commonly used.

Choice of mix. The table on p. 10 lists the standard mixes used in ordinary construction. Since these mixes meet all local building codes, as specified, you can rest assured that each one has many times more strength than is necessary for its intended purpose.

Therefore, there is no point in using a more expensive mix (because it has more concrete) when a less expensive mix will do the job. There is one exception, however: many masons opt for the high-sand mixes when making floors and walks because these mixtures are much easier to smoothen (finish) than the others.

TOOLS AND EQUIPMENT

If you hand-mix, you will need a bucket or similar measuring device, a large plastic barrel for water, a masonry hoe for mixing, a straight-edged shovel for removing the concrete and a mortar box for mixing. If you machine-mix, you will not need the mortar box and hoe, but you will need everything else, including a shovel of some kind. Also, you will require either a wheelbarrow or a mortar pan into which to dump the mortar from the machine.

Materials. Unless you have some special purpose, one type of cement is just as good and just as strong as another. Be careful not to use any cement bags marked MORTAR CEMENT, as these bags contain quantities of lime. Lime is never added to concrete, as it acts to weaken the concrete.

Gravel is graded by size and type. Small-size gravel is first choice for walks and floors, and where the concrete has to be pushed a considerable distance down or across a form. It moves easily and, being small, tends to sink into the mixture, making gravel concrete comparatively easier to finish. Larger-size crushed stone should be avoided, if possible. It is difficult to shovel, difficult to push along and a general nuisance on the job. However, if you plan to drop the concrete directly into the form, it

Fig. 9. *Three-quarter-inch crushed stone. As you can see, stones are not all alike in size, but few are much larger than their specified size.*

(7)

is no problem whatsoever. For all-around use, ¾-inch crushed stone (Fig. 9) is fine. Not as easy to move and finish as small-size gravel, it is still fairly simple to work with.

In answer to what sizes of stones provide the strongest concrete, be advised that a mixture of sizes is first choice because the multisize stones can pack together more tightly than stones of any uniform size. As to which makes better concrete, crushed stone or rounded gravel, the end result, using either, is identical. In any case, don't waste time or money searching for mixed-size stones or gravel rather than whatever the yard may supply. The differences in final concrete strengths, though they do exist, will be negligible.

As with sand, it is important to make certain the stone is perfectly clean. Commercial sources are always clean, and you need not concern yourself with those. If you dig your own from a nearby gravel bank, however, test the stones for cleanliness. Follow the same technique suggested for testing sand and water in Chapter 1.

Chemical change. As soon as water is added to cement, an irreversible chemical change begins, called hydration, which terminates with the cement permanently locking the sand and gravel in an ageless, stonelike bond. This is concrete in its elemental form. During hydration, the temperature increases, the mix stiffens and loses plasticity, and excess water within the mix is forced to the surface in a process called bleeding. The more or less simultaneous stiffening and bleeding occur at a point in time called initial set, or, simply, set. When the temperature of the mix, with the water added, is about 70° F, initial set follows in about one hour. At lower temperatures, the time lapse is greater; at higher, it is less. Excess water in the mix will slow set a little. Insufficient water will induce false set: the mixture stiffens almost as soon as mixing is stopped. In such cases, add more water and mix until proper lubrication is attained. Note that hydration will take place regardless of whether you are hand-mixing or using a machine, and even when the concrete is covered with water. The controlling factors are time and temperature.

All of this is very important. Up until set you can mush the concrete about, move it from place to place, and shovel and smoothen it. Following set, you have to break the concrete into pieces if you want to change its shape or shovel it from one spot to another.

Rough smoothing (screeding) can be accomplished up until initial set, as indicated by bleeding

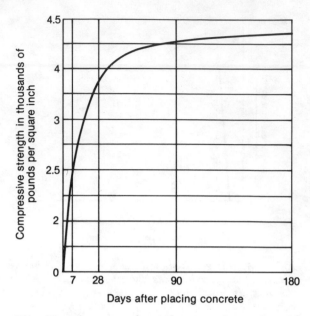

Fig. 10. The strength of concrete increases with time. Seven-day-old concrete is two and a half times stronger than one-day-old concrete.

or a general stiffening. Final smoothing (troweling) can be done shortly after set for ten to twenty minutes, or as long as the set concrete will respond. Ten or so hours later, the concrete will reach a condition known as final set. At this point it has become stone. Up until final set you can—if you must—break the concrete into powder, add a little water and reshape it to suit your purpose. Break the concrete up after final set, and you have nothing but small pieces of concrete: stones.

Curing. When concrete reaches a condition of final set, it is nevertheless still "green." It has a greenish color and is surprisingly weak and soft. Curing, the passage of time plus the absorption of water, is the process by which concrete grows harder as it ages (Fig. 10). Typically, concrete that will resist a compressive load of 1,500 psi (pounds per square inch) three days after pouring into position will resist over 2,000 psi after seven days, almost 4,000 psi after twenty-eight days, 5,000 psi at the end of three months and 5,500 psi after a year. Concrete must have sufficient water for proper cure. This can be accomplished by covering the concrete when it has reached initial set with a sheet of plastic, wet hay, wet newspapers or even wet straw. Wetting the concrete down periodically with a fine spray of water is also effective.

Once the concrete has reached final set, any

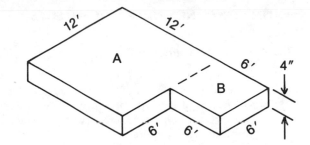

Fig. 11. To prevent a slab of fresh concrete from drying out while it is curing, cover it with wet newspapers. Plastic sheeting, straw or hay may also be used.

Fig. 12. To find the total concrete needed for this sample slab, divide it into two or more rectangular areas (A and B, as shown). Change 4 inches (slab thickness) to its decimal equivalent (0.33 feet). Find the volume of slab A (12 × 12 × 0.33 = 47.52 cu. ft.) and the volume of slab B (6 × 6 × 0.33 = 11.88 cu. ft.), then add them together to find the total volume (59.40 cu. ft.). Divide this figure by 27 to find cubic yards (59.40 ÷ 27 = 2.20 cu. yds.), then apply the waste factor (15% in this case) of 0.33 cubic yards for a grand total of 3.53 cubic yards. This is the working quantity of concrete you will need for the job.

amount of water in almost any force can be poured on. No harm will be done.

ESTIMATING REQUIRED QUANTITIES

Sidewalks, driveways and floors are called flat work. In such cases, the area to be covered is calculated first; this figure is then multiplied by the desired thickness. All measurements are made in feet and decimal fractions of feet. This results in a total cubic-foot figure (Fig. 12). It does not, however, account for the concrete that will stick to the tools, delivery chutes, wheelbarrows and the like. Neither does it account for the variations in slab depth that are normal for all dug-up slab bottoms. Generally, a waste factor of +20% is used when the required quantity is less than 2 cubic yards (a cubic yard is equal to 27 cubic feet); 15% extra is used when you need 2 to 5 cubic yards, and 10% when more than 5 cubic yards are needed.

When the slab is other than a simple square or rectangle, break it up into smaller squares and rectangles and add the areas of each before multiplying by thickness. When the shape is curved (for example, a free-form patio), duplicate the shape to scale on graph paper (Fig. 13). See how many boxes on the paper are fully occupied, then how many are partially occupied. Add the whole numbers and fractions and multiply by the desired thickness. In this way you can secure a fairly close estimate. Note that the resultant figure in cubic feet or yards is the

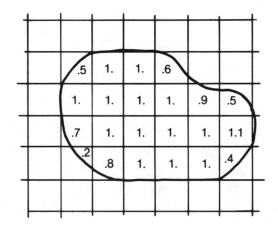

Fig. 13. To estimate the volume of concrete needed for a free-form slab, first draw the slab to scale on graph paper. Let each square equal 1 square foot, and estimate what fraction of each square is covered by the slab. Convert this fraction to a decimal and write the number in each square, as shown. Add the fractions (or whole numbers) for the total number of square feet—in this case, 18.7. Then multiply by the thickness.

quantity of concrete that you require: it is not the quantity of materials required to make the concrete. That figure, as you shall see, is considerably different.

ESTIMATING INGREDIENT QUANTITIES

CONCRETE MIX FORMULAS

1:2¼:3   This formula has the highest proportion of cement and sand, and is the strongest. Best for walks, driveways, floors and swimming pools. Used where concrete with minimum permeability is required.

1:2¾:4   Moderate strength. Good for foundations, cellar floors and similar construction not directly exposed to the elements.

1:3:5   The weakest of the three. Used for footings, massive walls, masonry support walls, abutments, etc.

More than meets the eye. If you trouble to check out any of the formula quantities, you will note the discrepancy in each. The numbers don't add up correctly: it takes about 1.5 cubic yards of sand, stone and cement plus nearly 5 cubic feet of water to make 1 cubic yard of concrete. Where does it go? As previously explained, the sand fits into the spaces between the stones, the cement fits into the spaces between the grains of sand, and the water fills the remaining spaces. All of this is merely interesting if you purchase your concrete ready mixed by the yard; if you purchase the ingredients separately, it is very important. If you don't use the figures given, you will not order sufficient quantity for your needs, and your cost estimates will be short by nearly half.

FORMULA QUANTITY TABLE

1:2¼:3 MIX

to make	1 cu. yd. concrete	2 cu. yds.	3 cu. yds.
cement	6 bags	12 bags	18 bags
sand	14 cu. ft.	28 cu. ft.	1.55 cu. yds.
stone	18 cu. ft.	1.32 cu. yds.	1.98 cu. yds.
water	36 gal.	72 gal.	108 gal.

1:2¾:4 MIX

to make	1 cu. yd.	2 cu. yds.	3 cu. yds.
cement	5 bags	10 bags	15 bags
sand	14 cu. ft.	24 cu. ft.	1.55 cu. yds.
stone	20 cu. ft.	1.48 cu. yds.	2.22 cu. yds.
water	30 gal.	60 gal.	90 gal.

1:3:5 MIX

to make	1 cu. yd.	2 cu. yds.	3 cu. yds.
cement	4.5 bags	9 bags	13.5 bags
sand	13 cu. ft.	1 cu. yd.	1.5 cu. yds.
stone	22 cu. ft.	1.6 cu. yds.	2.4 cu. yds.
water	27 gal.	54 gal.	82 gal.

(Note that the above formulae assume that the sand has bulked, that is, absorbed a quantity of water. This eliminates the complex problem of measuring the water present in the sand. At worst, there is a little more sand present than necessary, which is always good; never less sand than needed, which is very bad.)

READY MIX
OR MIX YOUR OWN?

There are two major sources of concrete: Ready Mix (sometimes called Transit Mix) and mixing your own. Which is best for you depends on a number of factors, and a thorough study is advised.

Selecting the wrong source can cost you lots of extra money, and possibly create lots of extra work.

Ready Mix. Very simply, you contact a yard and order the quantity you need. The yard will either suggest the mix acceptable to the building department or prepare what you desire. The ingredients are mixed dry en route in the rotating drum on the back of the truck, and water is added at your direction when the truck arrives. Generally, the cost of the concrete includes the truck's staying thirty minutes at the job site. Overtime carries an additional fee.

The machine does the mixing; mixing is thorough; water control is excellent; the concrete is delivered to the job and can even be poured right into your forms. Furthermore, the price is usually lower than if you mixed the same quantity yourself. The yard can also add whatever chemicals you may wish to improve water resistance, speed or slow setting or reduce concrete frost sensitivity. Concrete cannot be poured in freezing weather; freezing breaks concrete up into a kind of rocky powder.

The drawback to Ready Mix is that you pay a premium for quantities under a cubic yard or two, and the truck cannot go anywhere at anytime, as it weighs better than 35 tons. If it runs across a water-soaked lawn, it will ruin the lawn and break buried

Fig. 14. Ready Mix is often the most economical source of concrete. The trucks, however, are huge and cannot always reach the job site. This monster just fit between two trees. (Note the hydraulic lift beneath the concrete chute, and the chute extension.)

pipes. The chute on the truck is of limited length; some companies will rent a chute extension, while some will not bother.

The quantity of concrete that is economical by Ready Mix truck is far more than one man can handle. You need a crew to haul, place and spread several yards at a time, as in the making of a driveway or patio. (The exception, of course, would be when and where the truck could drop the concrete directly into a deep form. In such cases, all you need to do is direct the chute.) Thus, when you order Ready Mix you must be prepared with crew and tools to handle the delivery. And you must also realize that the chance of your securing the delivery at precisely the time you desire is very small. Your delivery may be the last in the day, which can make for additional difficulties.

When the terrain is such that the truck's chute cannot be brought to pour directly into the form, or when the delivery point is higher than the delivery truck, the concrete has to be hand-wheeled to the job. This is tough, back-breaking work. Concrete weighs 150 pounds per cubic foot. If the truck delivers 5 yards, let us say, that makes 135 trips. So you are stuck with the difficult choice of spending hours wheeling the "mud" (and paying for the truck's waiting time) or renting a dozen barrows and finding an equal number of stout lads or ladies to help wheel them. Actually, there is an expensive alternative, and it is most practical. Rent a concrete pumper (check with your masonry yard). This is a truck-mounted pump which can take the mud from the transit truck and pump it several hundred feet, via flexible hose, in any direction. Handling the hose end is a snap.

HAND MIXING

Hand mixing. Here you control the rate of concrete production. But unless there are several people working, the progress of the job is slowed down to a crawl. However, when working alone, this is the best method for most of us.

Hand mixing introduces the need for a place to store the materials: cement off the ground, sand and stone on a clean surface of some kind (otherwise the lowest layer has to be discarded because of the dirt). You will also need mixing tools (or a power mixer), and possibly a wheelbarrow to move the mud from mixing pan or machine to the job.

Hand mixing does not completely eliminate the time element. You can, for example, pour a portion of a walk, return the following day and pour the bal-

Fig. 15. Hand mixing is easy if you have the proper equipment. Formerly, all concrete was mixed this way.

Fig. 16. Using the "sandwich" system to ease mixing. Stone, sand and cement are evenly distributed in three layers. Then a little of each is scooped into one end of the box and mixed with an excess of water. The extra water makes mixing easier. Gradually, more "dry stuff" is added to bring the mix to the desired consistency. No attempt is ever made to mix more than a little at one time.

ance: the fresh concrete will bond without problem to the in-place concrete if not more than twenty-four hours pass. If more time passes, however, or the day is very hot and dry, the new concrete may not bond properly with the old and there will usually be a demarcation between the two. In some instances this is of no consequence; in others, the results may be unsatisfactory. There are, of course, ways in which you can hide the daily termination point and even extend the delay interminably, but these are not possible for all situations. In the case of a walk, for example, you can place expansion strips (explained later) across the walk at regular intervals: but on a large floor, you cannot very well quit for a couple of days with the job partly done.

Hand mixing is not very difficult if you have the proper equipment, follow the suggestions given, and do not try to do too much at one time.

Equipment. If you need more than a couple of cubic feet, don't try to mix the ingredients in a wheelbarrow or a mortar pan: secure a mortar box, which can be rented. Mortar boxes are rated by the quantity of mortar they can easily accommodate. Don't get an undersized box, as you will defeat your purpose. You will also need a mortar hoe, a straight-edged shovel, and some type of bucket or box for measuring. With care, you can use a barrow for measuring (the quantity is not important; it is the ratio that you are after). Additionally, you will need a large plastic garbage can and a small metal or plastic bucket for measuring the water.

Procedure. Start by wetting the inside of the box, then spread as much stone as you plan to mix at one time evenly over the bottom of the box. Cover this with the correct quantity of sand, and spread the sand evenly. Cover this with the correct quantity of cement. (Note that the quantity of sand given in the table is higher than that usually given in other tables. This is done to correct for bulking—when sand gets wet, it increases its bulk.) Next, pour a quantity of water into a corner of the box and, with your hoe, work a little of the three-layer mixture into the water. Just mix this little quantity of water with the sand, stone and cement in the corner. If the mud gets thick, add a little water; if too thin, slice some more of the three-layer mixture into the mud. When it is all one color, and not before, the little puddle of concrete is ready to be removed and used. Now, add water to the open space. Use the hoe to pull some more three-layer mixture into the water. Mix this batch until it is ready, and so on.

Since you are always adding a slice of the three-layer mixture, your ratio remains the same as you work your way across the box. Since you always have plenty of water in the batch you are mixing, mixing is easy and thorough.

Consistency. For most work, you want concrete that can be formed into a rough pyramidal shape. To test, dump a shovelful onto a clean board. If the mix slumps into a Sunday pancake, there is too much water. Simply pull some more dry stuff into the batch you are mixing. If the pile does not slump at all, you need a little more water. Bear in mind, however, at this point that half a glass of water is the difference between thick and soupy, so go easy on the water. This is why you should always work from a garbage can filled with water and a small bucket: you cannot judge how much water you are adding when using a hose the way you can with a bucket.

Clean your tools. Do not let the concrete dry on your tools or in the mortar box. If it does, it may adhere with more tenacity than glue. Hardened concrete on tools and box or pan just adds to the work.

MACHINE MIXING

When you need more than a few yards of concrete and you have help, or when you can position the mixer so that it dumps right into the form, a small power mixer becomes practical (Fig. 17). On smaller quantities practicality is a moot question. First there is the rental cost; second the need to secure and return the machine; followed by the need to clean the machine when you are finished.

Which machine. When selecting the machine, bear in mind that although a cubic-yard machine has a bucket of that capacity, it cannot hold or mix more than half this amount at any given time. Remember, too, that the small gas engines driving the mixers are

Fig. 17. A typical power mixer in the discharge position. Remember that the machine cannot handle more than half of its bucket capacity.

not always easy to start. Try yours before hauling it home from the rental store.

Procedure. Set the machine up on a level surface. Place rocks or bricks behind the wheels so that it won't move while operating. Start the machine, empty. Let it run until it is warmed up, then dump some water inside; follow with sand, stone and cement. Add a little less water than you believe necessary to complete the batch. Let the machine run until the concrete is one solid color. Dump a little out to check on consistency. If too stiff, add a little water.

Stand clear when adding ingredients. The machine must be turning over when you do this, as it cannot start loaded. Watch the shovel. If the blades catch it, the shovel will be ripped out of your hands and the pieces will fly. Pour what you need when you need it, but never stop the machine.

Do not leave the mud inside the machine more than thirty minutes or so—it may set right inside. When you have finished with the machine for the day, add clean sand and water and let it run for a few minutes, then dump the mix. This cleans the drum.

CHAPTER THREE

Brick

After stone, brick is the oldest building material used by man. Early bricks were made of clay, which was quite common in the Tigris-Euphrates basin, where the first bricks were supposedly made some 8,000 years ago. The clay was shaped with the aid of molds and was sun-dried. Our Western Plains Indians also made bricks like these, and many people still do for building homes. We call them adobe. Eventually, the early builders learned the value of fire-hardening their bricks, and by 560 B.C. fired and even glazed bricks (bricks with a porcelain surface) were common in the Near East.

Today, bricks are still made in the same basic way: clay is shaped in molds and placed in ovens where it may remain up to 100 hours. The chief difference is that crushed slate (a kind of stone) is added to the clay before it is shaped and fired. Also, nobody uses straw to make bricks anymore.

TYPES OF BRICKS

There are four basic types of bricks, which are manufactured in over 10,000 different sizes, shapes, surfaces and colors. One type of brick, however, called building brick (which is made in only two sizes and three grades), is used for an estimated 98% of all brick masonry work. So brick selection can be quite simple.

RELATIVE BRICK SIZES

(all dimensions in inches)

Standard	2¼ × 3¾ × 8
Modular	2¼ × 3⅝ × 7⅝
Jumbo	2¾ × 3¾ × 8
Norman	2¼ × 3⅝ × 11⅝
SCR	2⅛ × 5½ × 11⅝
Roman	1⅝ × 3⅝ × 11⅝
Baby Roman	1⅝ × 3⅝ × 7⅝
Fire Brick	2½ × 3⅝ × 9

Fig. 18. Oversize face brick was used for this beautiful circular garden wall, the patio it encloses and the adjoining walk.

Building brick. Always rectangular in shape and always the standard brick-red in color, this type of brick is the least expensive (Fig. 19). Its low cost is due to the large volume in which it is manufactured, the minimum amount of care taken in production, and the low heat applied in firing it. Length can vary plus or minus ¼ inch, which means there can be as much as a half-inch difference between two bricks in the same batch. Also, some of the bricks may be a little warped; however, they are strong and durable and, as stated, the least expensive of all.

Building brick is made in three grades: *NW,* non-weathering, which means it should only be used indoors; *MW,* moderate weathering, which limits its use to southern climates outdoors; *SW,* severe weathering, which can be used anywhere. Most ma-

Fig. 19. Building brick, the cheapest and the most widely used, varies somewhat in dimensions from brick to brick, and may also be a little warped and cracked.

sonry yards stock nothing but the SW bricks; they don't want to bother to stock three types. If you are planning to use the bricks indoors, however, it pays to ask for the NW bricks, as they are the least expensive of the three types, followed by MW.

As mentioned, building bricks are made in two sizes: they are called standard and modular. The standard building brick is 2¼″ × 3¾″ × 8″; the modular brick is 2¼″ × 3⅝″ × 7⅝″. Since many supply yards and many masons call both bricks common, and since brick size varies so much between bricks on the same pallet, you cannot simply measure one brick to be certain of what you have. You either have to check the size of several or ask the supplier.

Face brick. These bricks are made with more care. Their dimensions do not vary very much from brick to brick. They are harder, having been exposed to a higher temperature for a longer time. With rare exceptions, they are all SW grade. But most important they are designed to be seen; hence the name, face brick. They are made in a variety of dimensions, but most frequently in one of the sizes listed in the accompanying table. The facing surface of the brick may be smooth, striated or glazed in any color. Their cost may run to ten times that of building bricks.

Fire brick. These are specially designed to withstand high temperatures. They are used to line fireplaces and furnaces. They are white in color and must be laid up (set in place) with fire clay. Ordi-

nary mortar cannot be used, as it will crumble in the heat.

Paving brick. Shaped like a building brick, this brick is exceptionally strong and is designed to withstand vehicular traffic. It is rarely used for residential driveways or walks.

TOOLS AND EQUIPMENT FOR LAYING BRICK

Fig. 20. A small brick trowel.

Fig. 21. A joint tool.

Fig. 22. A brick set or mason's chisel.

Fig. 23. A small sledgehammer.

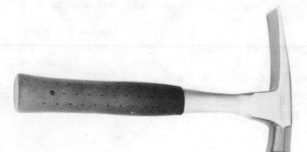

Fig. 24. A brick hammer.

You will need a mortar box or pan for mixing the necessary mortar, a hoe or shovel for mixing, a mortar pan or board on which to place the mortar while working, and a straight-edged shovel for moving the mortar from box to board. Also, a brick trowel, a joint tool, a stiff brush, and a small sledgehammer and brick set for cutting the brick. As an alternative to the sledge and set, you can use a brick hammer: its sharp end can also be used for cutting brick. Additionally, you will need a length of mason's line (string), a line level, (which is a very small level designed to be hung on a line) and a spirit level 30 inches or longer.

LAYING BRICK

Laying brick is not at all difficult. You can do a good strong job if you follow the instructions closely, take your time, work carefully and do not start your career as a bricklayer with a highly visible job, as for example a fireplace or veneering the front of a house. Start with work that is not directly in sight and work your way up to "face work," brick that faces the viewer directly.

Brick preparation. For proper adhesion between mortar and brick, the bricks should contain a little moisture. A bone-dry brick draws water out of the mortar and weakens it; a dripping-wet brick tends to slide out of place. Check moisture content by placing a few drops of water on the brick. If the water disappears in a minute or two, the brick is too dry. Put the garden hose to the pile and let the water flow until the bricks will not absorb anymore. Then let an hour or so pass and they will be ready to lay up.

Footings. In the trade, the surface upon which brick or any other masonry rests is called the footing

or support. Brick cannot be laid on soil or on wood; in either case, the resultant wall will eventually crack. Brick can be laid on a concrete floor or driveway when the brick supports nothing but itself, as in the case of a wall or a planter, but when the brick is to support a building, the footing must conform to building codes. In any case, brick is always laid on a solid concrete or stone base. Footings and foundations are covered completely in Chapter 11.

Laying the first course. A course is a row or layer of bricks or stones or blocks. The first course is the lowest, the course that rests on the footing.

Let us assume that we are going to build a wall a single brick thick and it is going to be exactly 8 feet long end to end using standard brick, which is 8 inches in length. We cannot lay the bricks against each other, but must place mortar between them: this is called a joint. Let us make each joint roughly ½ inch thick. Thus, if each brick is effectively 8½ inches long, how many bricks will we need for the first course?

If we divide 96 by 8 inches, the length of the brick, we come up with an answer of 12 bricks even. But this is no good since 12 bricks does not leave us with spaces for the joints.

Let's try 11 bricks.

$11 \times 8 = 88$

$96 - 8 = 8$

If we use 11 bricks we have 8 inches for 10 joints. (There are only 10 joints between 11 bricks.) Dividing the 8 inches into 10 parts gives us .8 inch; much too much for a joint.

One solution is to use 11½ bricks. This number of bricks alone totals 92 inches, leaving 4 inches for 10 joints.

$4 \div 10 = .4$ inch. This is a suitable joint width.

Thus for our desired wall length of 96 inches, using 8-inch-long bricks, we can do the job with 11½ bricks and .4-inch mortar joints between them.

Is there an easier way?

Yes: vary the wall length to suit the bricks and joints. In this case, the wall might be extended to 101½ inches or shortened to 93 inches, using whole bricks and ½-inch joints.

Now, knowing the number of bricks that will be used to make the first course, the joint spacing between the bricks, and the desired wall length, we are almost ready to lay brick. First, as mentioned, we need to make certain the bricks are moist. Second, we need to prepare sufficient mortar for an hour or so of work. This is accomplished by placing the mortar (prepared as in Chapter 1) on a mortar board or in a mortar pan close by the job. To work efficiently, you want to be able to reach the mortar without taking more than one step. Similarly, the bricks must also be within easy reach. Now we are ready to begin.

Fig. 25. A small footing has been poured and is now being screeded (this is to be used for the brick-laying "demonstration"; for a permanent support, its surface would have been made several inches lower).

Mark the beginning and end of the wall you wish to erect on the footing. With a straight edge, mark the desired front edge of the entire wall (Fig. 26).

Fig. 26. A nail has been driven into the fresh concrete: this marks one corner of the job. A second nail locates the second corner. Now the line that will demark the front edge of the wall is scratched into the surface of the concrete.

Scoop a quantity of mortar onto your trowel and dump it on the footing so that it is near one end of the wall, as marked. Spread the mortar to form a flat mound a little wider and a little longer than the first brick, about ¾ to 1 inch high (Fig. 27). Place

Fig. 27. A glob of mortar is spread over the footing into a flat mound a little longer and wider than the brick, and about ¾ to 1 inch high.

the first brick on the mortar, flat side down (as opposed to on edge). Position it so that one end is exactly in line with the end mark of the wall, and its side forms the beginning of the front line of the wall (Fig. 28). The front line may have become covered with mortar, so you will have to judge this accordingly.

Fig. 28. Press the first brick down until it is ½ to ¾ inch above the footing. Its front edge should be just above the scratched line, its end just touching the nail.

Press the brick down firmly until it is ½ to ¾ of an inch above the footing. Now, with the aid of a spirit level, make the brick perfectly level (bubble in the center) end to end and side to side. Use the heel of the trowel to help you (Fig. 29).

Fig. 29. Place the spirit level atop the length of the brick. With the help of the hammer handle, adjust the brick until the bubble is centered. Repeat the operation with the level positioned across the width of the brick.

If you find you cannot move the brick easily, the mortar is too dry. Remove brick and mortar, add a drop or two of water to the whole batch, remix and try again. If you find the brick slips off, the mortar is too wet; add some cement, mix it up and try again.

You should now have one end brick perfectly level in two directions and exactly in line with the front and end of the wall-to-be. Next, go to the other end of the wall line and position a second brick there in the same manner as the first (Fig. 30). This may be a whole brick or a half brick, based on your earlier calculations.

Cutting brick. Hold the brick in one hand and rap it smartly with the blade of a brick hammer. It should break easily and evenly. Or place the brick on the ground or on a board (never on concrete). Position the edge of a bricklayer's set (broad-blade

mason's chisel) across the brick. Rap the set's end with a small sledgehammer (this works better than an ordinary hammer) and, presto, the brick is cut (Figs. 31 and 32).

Fig. 30. Repeat the leveling operations at the far end of the wall using half brick.

Fig. 31. Cutting a brick with the point of brick hammer.

Fig. 32. Cutting a brick with a brick set and small sledge.

Figs. 33, 34. With the second brick in place, use two additional bricks as weights and stretch a string from the top of one brick's front edge to that of the second brick. Adjust the bricks so that their front edges are just clear and parallel to the string.

With the second brick in place, use two additional bricks as weights and stretch a mason's line (string) from the top of one brick's front edge to that of the second brick. Adjust the line so that it just misses the front edges of the bricks (Figs. 33 and 34). Next, using the line as your guide, adjust the two in-place bricks so their front edges are parallel with the line. Then recheck both bricks with the level to make certain they are still level end to end and side to side. If it is important that both ends of the wall be exactly the same height, hang a line level from the center of the string between the two bricks. If the line level's bubble is centered, the two bricks are at the same height; if not, push the higher one down until the bubble is centered. Naturally, you will have to check the level of the brick you adjust again, end to end and side to side.

At this point, you have the two end bricks in place. Spread a bed of mortar on the footing along the wall's front line from one brick end for a distance of 8 or so inches. Butter the end of a brick (force some mortar to adhere to its end) with enough mortar to make a joint thickness equal to your established needs (Fig. 35). Place the buttered brick atop the mortar bed and gently slide it (buttered end first) into place against the end brick (Fig. 36). With the aid of your hand and/or the heel of the trowel, make this brick level in two directions, and its front edge in line with the adjoining brick. Repeat this operation at the other end of the wall.

Fig. 35. Butter the end of a brick by forcing mortar against it.

Fig. 37. Brick by brick, the space between the two in-place end bricks can now be filled in.

Fig. 36. Since we are going to construct a corner in this "demonstration," this buttered brick is placed on a bed of mortar and at right angles to an end brick (the angle has been marked on the footing). Note that the string has been temporarily removed.

Continue adding bricks to the wall alternately at both ends until you have space for only one brick more. This is called the closure brick. Butter both ends and gently lower it into place. If you press the mortar hard against a brick's end, it will stick; all it takes is a little practice (Fig. 38).

To lay the second course, start with a half brick laid over the whole brick on one end. If there is a whole brick at the other end, put a half brick over this one too; otherwise put a whole brick on top of the half. Take care not only to keep the brick perfectly level in two directions, using the level as before, but to keep it in perfect vertical alignment with the brick below. Do this by using the spirit level in a vertical position against the sides of the end brick (Fig. 41), and repeat the operation at the other end of the wall. Fill in the balance of the course with the required bricks. Or, do the ends first for several courses and then fill in the middle. The third course starts and ends with a whole brick; the fourth with half bricks and so on up.

As you work, take care to keep the joints between courses of equal thickness so that the top surface of the wall remains level and looks attractive. But note, so long as you use a good mortar with sufficient water and the bricks are a little moist, the wall will be strong whether the joints are even and attractive or not.

Fig. 38. *The center, or closure, brick is now positioned. Both of its ends must be buttered. To hold the mortar in place against one brick end, the trowel is sometimes used.*

Fig. 39. *With the first course completed (or left for later completion), a new bed of mortar is laid down on the end and side bricks.*

Fig. 40. *A brick is placed on this new bed of mortar (note that this brick overlaps two lower bricks).*

Fig. 41. *The spirit level is used to make certain this brick is in perfect vertical alignment with the lower bricks.*

Fig. 42. *The next brick is laid in place. The guide string is stretched to the other end of the wall as before, and more bricks are laid.*

Fig. 43. As the wall is erected, the edge of the trowel is used to keep it clean.

Fig. 44. The joints are then tooled (pointed) or finished, with a joint tool.

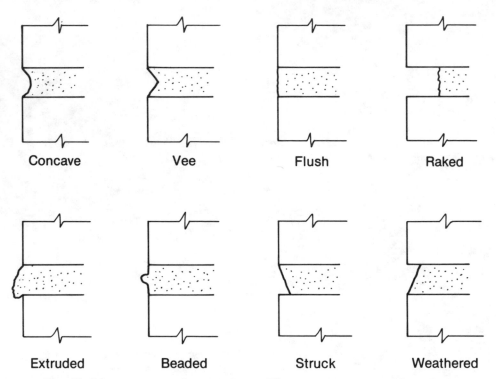

Concave   Vee   Flush   Raked

Extruded   Beaded   Struck   Weathered

Fig. 45. Common types of mortar joints. The concave joint is made with the joint tool shown in Fig. 21. Struck and weathered joints may also be made with the point of a trowel.

Keep it clean. When you have laid up three or more courses, take the edge of your trowel and scrape it over the joints so that no mortar protrudes; then clean up whatever mortar has fallen onto the footing (Fig. 43). If mortar has stained some of the bricks, brush them as clean as you can with a stiff brush.

Finishing. To improve the overall appearance of the job, and to strengthen the joints they are tooled (some call it pointing). Very simply a joint tool is pressed against the joints and slid over them at the same time (Fig. 44). This compresses the mortar into the joint and makes its surface smooth. The shape of the tool controls the shape of the finished joint. If you wish, you can use the point of your trowel for tooling, or a short length of ½-inch tubing.

(22)

CHAPTER FOUR
Concrete Block

Concrete block is made of concrete molded into various block shapes, cured by subjection to steam, and then aged for a short time before using. They were first made sometime early in this century, but did not find wide application until fairly recent years. At first, their use was more or less limited to foundations, replacing the rubblestone foundations popular years ago. As the price of lumber increased, an increasing portion of the entire building was made of block. Today, its use is still increasing; if and when earth-sheltered homes (homes that are almost entirely underground) become commonplace, most of them, as well as most commercial buildings, will be made of block.

TYPES, SHAPES AND DIMENSIONS

There are two types of blocks. One is made from crushed stone, sand and cement; the other is made from either cinders or a lightweight volcanic rock, sand and cement. The first type is sometimes called standard, but more often simply concrete block. The second type is called lightweight block, and weighs about half as much as an equal-size standard block.

Lightweight blocks are much more porous than standard blocks and are not suited to wet applications such as cellar and basement walls. They are also not as strong as standard blocks. For these two reasons, many building departments limit their use to light-load, aboveground structures such as cabanas, garages, hen houses and partition walls. From a mason's point of view, the lightweights are much easier to work with, simply because they are lighter. In many areas they are also less expensive.

Lightweight blocks are manufactured in perhaps a dozen different shapes. Standard blocks are manufactured in possibly a thousand different shapes and sizes. This is nice, but rarely helpful. Blocks are heavy and bulky, and it is easier and cheaper to transport their raw materials than the blocks themselves. Therefore, you will find block makers spread all over the map. Since no single block company makes every type of block, your choice more or less depends on what is manufactured locally—unless you don't mind paying freight charges for your blocks.

Common shapes. Three basic block shapes account for better than 95% of all blocks made and used (Fig. 46). The reason is simply that these three shapes, in a limited number of sizes, are all you need to build 95% or more of everything that can be made of concrete block.

Two shapes have core holes, and one is solid with no holes. The cored blocks are rectangular and are always 7⅝ inches high. They may have two or three holes. The three-holer is a bit heavier, stronger and more expensive than the two-holer. For ordinary construction, one is as good as the other, but note that some building departments insist on the two-holer. The solid blocks are called 4-inch and 6-inch solids, respectively.

The most common cored block is called a stretcher. It has indentations at both ends and is made for use between the ends of a wall. If you place it at the end of a wall, you have to fill the indentation with mortar to make it flat.

A variation of the cored block is called an end block. It has an indentation at one end and a flat surface at the other. It is usually used for the ends of walls, but may be used anywhere if you wish. Double-ended blocks are also manufactured. They have two flat ends.

Uncommon shapes. These account for all the other different block shapes manufactured. A number are made to simplify masonry construction, as for example, concrete bricks, half blocks, double-

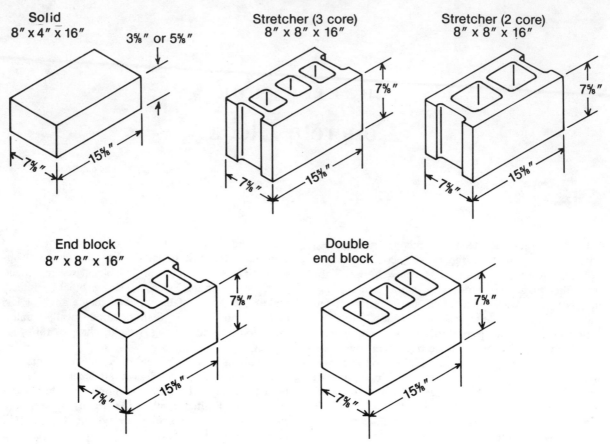

Fig. 46. Only a few types of blocks are used for 95% of all masonry-block work in this country.

ended blocks, partition blocks, chimney blocks, etc., while others are made for decorative purposes.

A concrete brick is a solid block of concrete the shape and size of a building brick. It is generally used as a filler where the color of building brick would make the brick objectionable. Sometimes it is used for ornamentation. A half block saves the mason the trouble of cutting a block in half. A double-ended block has a flat surface at both ends, as mentioned earlier. It is used for making piers (pillars), among other things. A partition block is a thin block with core holes, and is used for making partitions which need to support only themselves. A chimney block is a large square block with a hole in its center. These blocks are stacked one atop another to make a chimney. The flue pipe goes down the center. One chimney block replaces four ordinary blocks.

There are also special blocks that hold windows and doors in place. They eliminate the need for shaping these portions of the wall out of mortar, but they also introduce the need for accuracy. If these special window and door blocks are not positioned accurately, the window or door will not fit. On the other hand, if you use mortar to hold doors and windows in place, you can correct for errors of an inch or two without difficulty.

Finally, there are ornamental blocks. Some have serrated surfaces, some have glazed surfaces, and others pierced and otherwise exotic shapes. Some are used for building walls, some for garden walls, walks, patios, etc.: the variations are almost endless.

Dimensions. Blocks are almost always designated by whole numbers. It is simply easier. When the actual dimension is a fraction short, the next whole number is used. In almost all instances, the difference between the block's actual dimension and its nominal dimension is 3/8 inch, as in stretcher and

Fig. 47. A very attractive but very difficult block wall to make. All the joints must be perfect or the results look poor, though strength is not affected.

end blocks, which are always 7⅝ inches high. Their lengths may be 16 or 18 inches nominally. Actually the blocks will be either 15⅝ or 17⅝ inches. Their front-to-rear dimension, again nominally may be 4, 6, 8, 10 or 12 inches. Again, the actual front-to-back dimension will be ⅜ inch less.

The reason for all three dimensions of blocks being ⅜ inch less than nominal is that space is left for mortar. Thus, if you build a ten-course-high wall, using ⅜-inch-thick joints, you will have an 80-inch-high wall. This makes for easy computation. The same goes for wall length. To make a 15-foot-long wall, you need just ten 18-inch blocks.

BLOCK DIMENSIONS

Shape	Nominal	Actual
Solid	4-inch	3⅝ × 7⅝ × 15 or 17⅝
	6-inch	5⅝ × " "
End or	6-inch	5⅝ × " "
stretcher	8-inch	7⅝ × " "
	10-inch	9⅝ × " "
	12-inch	11⅝ × " "

Note that only one dimension differs on all these blocks. This is done so that they can be easily used together.

TOOLS AND EQUIPMENT

You will need the same tools and equipment as listed for working with brick.

LAYING BLOCK

So far as skill is concerned, less is required to lay block than to lay brick. For one thing, there are fewer blocks to lay to make a given wall area; for another, the very size of the blocks makes it much easier to keep them perpendicular and in line. Block, however, weighs more than brick, so when laying 10-inch or larger blocks it is advisable to have two people on the job. Not that one person cannot lift the blocks; it is just that you have to ease the blocks gently into place, and this is where an extra pair of hands is useful.

Block preparation. Unless you want to lay block immediately following a heavy downpour, your supply of block is normally left to the weather, unprotected by a plastic sheet or anything else. Conversely, unless you are working in desert country, block is not usually moistened prior to laying up. But if the block stiffens the mortar too quickly, as it will when bone dry, hose the block down before using.

Footing. Like brick, concrete block must be firmly supported by solid masonry. If the block is to carry a load—if, for example, it is to be a house foundation—a building footing beneath the block is required. This is discussed in Chapter 11. A non-load-bearing partition, however, or a dividing wall, such as might be used to add charm to a basement playroom, can be erected on a concrete floor.

Planning the first course. If possible, select a wall length that works out to an exact multiple of the *nominal* length of the block you plan to use. Hold mortar joints to ⅜ inch, and you will end up on the mark. If that is impossible, try various joint widths up to ⅝ inch maximum and ¼ inch minimum, and see how that works out on paper. If varying joint width and block length doesn't work out for you and you must hold to a specified wall length, you'll have to cut a block in each course. This cut block, called the closure block because it is placed last, should be cut to fit the remaining opening. In this way, variations in joint width can be accounted for with the last block in the course.

Cutting block. This can be done with a brick hammer or a brick set. In either case, mark the desired line of cut on both sides of the block. If you are using a brick hammer, hold the block at an angle with its lower end resting on earth or a plank—never on solid masonry. With the point of the hammer, strike the block repeatedly along the desired line of cut until you have made a slight groove all across the block. Then turn it over and do the same along the desired line on that side of the block. If the block doesn't crack, turn it over again and keep striking it in the groove until it does. If the block cracked in the wrong place, it was defective; don't let anyone tell you you lost patience and struck the block too hard.

To use the brick set, place the block flat on the ground or a plank. Place the edge of the set on the desired line and strike the set lightly with a hammer while you move it along the line. When you have made a slight groove on one side of the block, turn it over and repeat the grooving operation on the second side. Sooner or later, the block will crack neatly along the desired line.

Laying the first course. Start by readying the mortar you will need and placing it on a mortar board or in a mortar pan convenient to the job. Place the blocks you are going to use for the first course handy to their ultimate positions. As with bricks, mark the start and end of the wall on the footing. With a straight edge, mark the desired front edge of the wall.

Scoop a quantity of mortar up onto your trowel and dump it on the footing near one end of the wall-to-be. Spread the mortar to form a flat mound, a little wider and longer than the width of the block, and about 1½ inches high (Fig. 49). Position an end block (flat end in line with wall end) atop the mortar, taking care to align the block with the end and front line marks you have made (Fig. 50). Since the front line may be covered with mortar, you have to use your eagle eye here. Small ends of core holes are on top.

Press the block down, keeping it as level as you can until its bottom rests about ½ to 1 inch above the footing. With the aid of a spirit level 30 inches or longer, make the block perfectly level side to side and end to end. A few taps with the heel of your trowel or the heel of a brick hammer will help you (Fig. 51).

If you find you cannot move the block, the mortar is too stiff. Remove block and mortar, add a little water to the mortar, mix thoroughly and try again. If the block slips out of position, the mortar is too wet: add a little cement, mix and try again.

Fig. 48. When you have to move block any distance, use a wheelbarrow and run the barrow over a plank road.

Fig. 49. A nail marks the desired corner of the wall, while lines scratched in the concrete footing show the front edges of the wall. As always, the first step is to lay down a bed of mortar, longer and wider than the first block.

Fig. 52. Since a corner is to be constructed, the second end block is positioned at right angles to the first.

Fig. 50. The first block is positioned atop the footing in line with the scratch mark and the corner nail.

Fig. 53. A mason's line is stretched between the two blocks, using bricks to hold the line parallel to the front edges. The blocks are then adjusted so that the front edges are truly parallel with the line.

Fig. 51. With the aid of a spirit level, the block is made level in two directions: end to end and front to back.

You now have one end block in position and perfectly level in two directions. Go to the other end of the wall line and position a second end block there (Fig. 52).

With the second block in place, use two bricks as weights and stretch a mason's line from the top of one block's front edge to that of the second block (Fig. 53). Position the line so that it just misses the

front edges of the two blocks. Next, using the line as your guide, adjust the two blocks so that their front edges are parallel to the line. Then recheck both blocks to make certain they are still level side to side and end to end. If they are not, adjust as necessary.

As with bricks, if it is important that the wall's top be perfectly horizontal (which means both end blocks must be at the same elevation), hang a line level from the center of the line now stretched from block to block. If the line level's bubble is not centered, push the higher block down until the bubble is centered, but do not push so much that there is less than ¼ inch of mortar beneath the block. Should that be about to happen, stop pushing and raise the other block. After this is accomplished, check the levelness of the block you have repositioned.

Next you have to fill in the space between the two in-place end blocks, with a number of stretcher blocks. Start by laying down a 1- to 1½-inch-thick bed of mortar following either end block. Stand a stretcher block on end and butter the ends with 1 inch or more of mortar (Fig. 54). Gently lower the block into place and slide it against the in-place block. Move it forward or backward as necessary so that it is in line with the mason's line, which is still in position (Fig. 55). Place the block so that there is a ⅜- or ½-inch mortar joint between it and its neighbor, depending on the joint thickness you want.

Now, with the aid of your spirit level, make certain the second block is also level in both directions and its top surface is flush with the adjoining block. Continue laying block until you have space for only one block or less. This is the closure block; generally masons lay block from both ends of the wall so that the closure block is somewhere in the center.

Measure the remaining space. If it will take a full block and mortar, fine. If not, you will have to cut the block. In any case, stand the whole or cut block on end and butter its flanges (protruding ends). Next, butter the flanges of one in-place block. Now you have to carefully lower the closure block into place. If the mortar falls out, let the block be. Place a stick or a gloved hand on the inside of the open joint and force mortar into the joint with the aid of the trowel: this is messy but strong. After using the level on this last block, remove the line, line level and supporting bricks.

Before you lay the second and subsequent courses, a little nomenclature is in order: the front and rear sides of the block are called its face shells. Those portions of the block that connect its face shells are called its webs (a three-hole block will have four webs). As mentioned before, the projecting ends of a block are called flanges.

As with brick, if the first course starts with a whole block, the second must start with a half block so that no joint is directly above another. If necessary, cut a block in half. Before you position it, place a 1-inch-thick layer of mortar on the outside end web of the in-place block and on top of both face shells (Fig. 58). Note that with the exception of the outside (nearest end) web of the two blocks forming the ends of the wall, it is not necessary, cus-

Fig. 54. The end of a block is buttered.

Fig. 55. A bed of mortar is prepared and the block is positioned. Note the small space that will be left open for the closure block.

(28)

Fig. 56. How a block is cut with a brick hammer.

Fig. 58. A bed of mortar is laid atop the corner of the wall and the first, second-course block is laid in place.

Fig. 57. The ends of the closure block are buttered and the block is slid into place. If the mortar falls out while the cut block is positioned, force the mortar back in place with your trowel.

tomary or required to place mortar on any of the other webs.

Place the first or end block of the second course (be it a whole block or a fraction of a block) directly on top of the in-place end block (again, you want an end block for the end of the wall). Press this block down until it looks level and is directly above the lower block, and the mortar joint is close to what you want. With the spirit level, make certain the block is level in both directions and that it is also directly above the lower block (Fig. 59). You may have to clean the joint of excess mortar to place the side of the level firmly against both blocks. When this is done, repeat the operation with another block at the other end of the wall.

Next, install the line from the top of one second-course end block to the other. Adjust the end blocks as necessary to make certain their face shells are parallel. Hang the line level from the center of the line to make certain their elevations (heights) are equal. Correct as necessary and then go back and check for verticality and levelness of both end blocks.

The third course begins and ends with whole blocks. The fourth course begins and ends with half blocks, and so on up. However, most masons build

Fig. 59. The spirit level is used to make certain the second-course end block is directly above the first-course end block. Joints may have to be cleaned to permit the level to make proper contact with the block.

Fig. 61. Third-course end block is positioned (note how blocks overlap). Mason's line and level will be used to make certain this block is parallel with other end blocks and vertical to lower blocks.

Fig. 60. Mortar is being spread to support a third-course end block.

the ends of a wall several courses high before they fill in the middle sections. It is a little easier this way.

Keep it clean. So soon as you have laid up three or four courses, take the edge of your trowel and scrape it over all the joints, discarding the mortar so collected. Then scrape the top of the footing free of fallen mortar.

Finishing. To strengthen and improve the appearance of the joints, the joints are tooled. This may be done by running the point of the trowel down the length of the joints, or doing the same with a joint tool. Either action presses the mortar firmly into the joints, insuring good adhesion and smoothing their surface. If some mortar is missing from a joint, simply press additional mortar into place. The surface of the joint depends on the shape of the tool used, and a variety of joint tool shapes are available: Vs, half rounds, etc. As an alternative, you can also use a short piece of small-diameter tubing, as mentioned previously.

Fig. 62. Joints are cleaned up with edge of trowel.

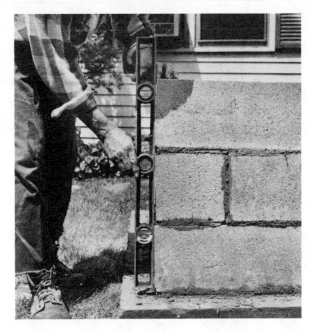

Fig. 63. A final check is made on third-course end blocks before proceeding. If all is well, it is advisable to tool completed joints before going farther.

CHAPTER FIVE

Stone

Stone is a lot easier to work with than most non-masons realize. While it is true that it is heavy and hard to cut, it is also true that it is an almost can't-fail material. No matter how crude the workmanship, the final result always has charm—that is the nature of stone. Even a rough stone wall, which need not be more than a long pile of rocks, has a beauty that cannot be duplicated by any other masonry material. Cutting, which is difficult, can generally be eliminated by simply selecting stones of the correct size and shape: when this doesn't work, a little trimming usually produces all the dimensional change necessary. When true cutting is required, the work and skill can be supplied by the quarry from which you purchase the stones.

Although stones are heavy, there is no reason you cannot roll the big ones into place and find a friend to help you lift them when necessary. Also, there is no reason to work at anything other than a slow and easy pace. And there is no reason at all why you must work with stones that are difficult to lift. Small stones do just as well.

KINDS OF STONES

For the purpose of this discussion, we can place all stones into one of four groups: quarry stones, fieldstones, flagstone and slate.

Quarry stone. This group includes any kind of stone quarried for building purposes, with the exclusion of flagstone and slate, which are also quarried. The stone may be plutonic or metamorphic, schist, gneiss, sandstone or granite. In addition to providing the stone, the quarry cuts it. Note that all stones are laid up (positioned and mortared in place) in exactly the same way regardless of the material from which they were formed.

The cheapest kind of quarry stone is called rubble. These are the odd-shaped pieces resulting from cutting and blasting. They are good for fill, rough walls and use with other, better-shaped (for our purpose) stones. The next step up in usefulness and cost is usually called *select*. These are stones that are selected for flatness and regularity of shape. Next would be partially trimmed, followed by fully trimmed or finished stones.

The first two grades are measured by bulk. They are dumped into the back of a truck and you pay for the overall volume of the load. In any given load, there may be 30% to 50% air: this air is also included in your bill. A truck with a 10-yard capacity would be considered to be carrying 10 yards of stone when its cargo space was filled to the top.

The next grade is often measured by square footage. The stones which are reasonably parallel-sided are laid on the ground, and the area they cover is measured. Charges are based on the area. Generally, such stones are between 4 and 5 inches thick. Fully trimmed stones are usually sold by the piece, so much per stone.

Each quarry has its own rates and method of computation. Up until the very top grades, which are often packed in wood and shipped cross-country, there is no simple way of determining what you will receive and how much it will cost without going to the quarry yourself. Top grade stones (meaning the care and accuracy with which they are cut and possibly polished) can be purchased through local mason supply yards.

Fieldstones. Any stone you can pick up for free, excepting stones found in someone's quarry, is called a fieldstone. Depending on your location, these stones can vary in size, shape and type. With luck you may find what you need in your back yard. With a little less luck and more effort, you can possi-

Fig. 64. Measuring veneer-grade quarry stone. These are selected stones having more or less parallel faces and are about 4 to 5 inches thick.

bly find what you need at construction sites, abandoned fields and demolition sites. Old houses once used rubble masonry for foundations, and these are easily pried apart.

Flagstone. This and bluestone are quarried. Both are forms of sandstone which naturally occur in layers. The bluestone is a little harder and blue-gray in color; flagstone comes in brown, pale red and shades of gray. Both flagstone and bluestone are sold by thickness and square footage, the 1-inch and 2-inch thicknesses are the most common. The 1-inch stones will vary from ¾ to 1½ inches in thickness, while the 2-inch stones will vary from 1¾ to 2½ inches. The prices of these two types of stones also vary with the method used for cutting. Snap-cut (which leaves a mildly serrated edge) costs less than sawn (which leaves an almost polished edge). Snap-cut is more or less standard. This is what all the yards and garden supply shops carry. You'll find the yards charge less and stock a greater variety of sizes.

Flagstone and bluestone can be used by themselves for walks, patios and the like, and they can also be used for flagging, which means placing them on top of concrete walks and patios, held in place with mortar.

Slate. Slate is a very hard form of sandstone, which also is found in layers. Its colors vary from al-

most blue through light gray, gray-blue and a reddish purple. Naturally, the red costs more: it is the more attractive. Slate can be had with a rough, wavy surface or a flat, smooth surface. It can be had with cut edges, which are always sawn, or with broken edges. There are no standard thicknesses to commercial slate. You have either to select your own or ask. Generally, masonry slate is stocked in thicknesses of about ¾ inch when it is flat and smooth; rough or wavy-surfaced slate can vary in thickness from ½ inch to 3 inches—in the same piece.

Slate can be used for walks and patios without support when it is thick enough, and it can also be used to flag walks and patios just like flagstone. However, slate is slippery when wet and should not be used on steps or steeply pitched walks. Slate is often sawn to specific dimensions for use as window- and doorsills; for such applications its surface is usually machine polished. Like flagstone, slate is sold by thickness, surface quality, size and method of cutting. And, as stated, color too has an effect on selling price.

TOOLS AND EQUIPMENT FOR CUTTING STONE

You will need a mason's chisel and small sledgehammer. If you plan to tackle boulders, you will probably find a full-size sledge necessary. You should wear gloves and safety glasses when cutting the stone.

CUTTING

Cutting quarry stone. Actually, we don't cut quarry stone, in the full sense of the word. What we do is split or trim. In either case, the stone must be placed on a firm but slightly yielding surface. Generally, the stone is placed directly on the earth or several inches of sand in a strong, raised box of wood. This arrangement eliminates the need for bending down. Sometimes, the stone is placed on a thick wood plank on top of a solid masonry support. The stone is never placed directly on concrete or any other solid masonry, as it will bounce and may shatter.

To split a stone, hose the stone down with water if it is dirty. Look for its grain and possibly an existing crack: this is the line along which the stone may split. It doesn't always, even for Italian masons. Position the stone with the grain or crack uppermost. If necessary, place some dirt or sand alongside to hold

Fig. 65. Cutting quarry or fieldstone with a small sledgehammer and mason's chisel. The trick is to locate a flaw or the grain so that you can secure an even-edged cut.

Fig. 66. Cutting flagstone. One side of the stone is grooved along the desired line, and then the other side is similarly grooved. This is repeated until the stone separates or you can break it off by just pushing down on the end.

the stone in position. Place the chisel on the grain line or crack, and rap the chisel with the hammer (Fig. 65). If a few strokes don't split the stone, move the chisel along the expected split line and continue punishing it. If that doesn't do it, try striking the stone along the expected split line with the large sledge alone (just be careful you don't hit yourself). If a dozen strokes of your sledgehammer don't produce results, or if the stone did not split where you wanted it to, follow standard masonry procedure: try another stone.

To trim a stone, place the stone flat on its support. If the lower side is convex, dig a little hole to accommodate the bulge. Place the chisel about ½ to 1 inch from the edge of the stone and rap it sharply with the small sledgehammer. Do not try to trim off more than an inch at a time.

Cutting fieldstone. The same technique may be used for splitting and trimming fieldstone. However, if you start with a boulder (a round stone), you will save time by going to the full-size sledge first. If you cannot see a grain line and the boulder doesn't split,

turn it on its side and try again. The grain may be invisible.

Cutting flagstone and bluestone. Mark the desired line of cut with chalk or blue pencil. Place a 2 × 4 or similar board on the ground and position the stone atop the board so that the line of cut is centered over the length of the board. Now, with a mason's chisel and a small sledgehammer, cut a shallow groove along the line. Then turn the stone over and cut a second groove exactly above the first. When you have done this, deepen the grooves on both sides with repeated light blows until the stone snaps along the groove. Just be careful not to strike the stone too hard at any one point. Or, slide the groove past the side of the supporting piece of lumber. Pressing down on the end of the stone should cause it to snap free if the grooves on both sides are deep enough.

Note that you can purchase precut flagstone in a variety of sizes at the mason's supply yard. Should you want special sizes or shapes, they can have the quarry cut the stone to your sketch—for a fee, of course.

Fig. 67. Cutting slate. Slate is not cut the same way as flagstone, but is crushed along the desired line. This method wastes a lot of slate, but the only other alternative is a power saw with a stone-cutting blade, which is neither inexpensive nor easy to use.

Cutting slate. Slate is neither cut nor trimmed: it is broken. The edge of the slate is pressed at an angle against a concrete surface, and the edge is mashed with a hammer. This shortens the piece of slate by an inch or less. The debris is brushed away and the mashing is repeated as often as necessary. Note that this method not only wastes a great deal of slate, but leaves broken, ragged edges. If you want to limit slate waste, take care in selecting the pieces of slate you plan to use. If you need straight-edged slate of a specific size, you will have to purchase sawn slate or have the slate sawn for you by the quarry or yard. It is next to impossible to saw slate with any hand-held power tool.

CHAPTER SIX
Walks

There are dozens of different types of walks. All of them can be made without difficulty by anyone even only slightly accustomed to using tools. Compared to all other masonry work, walks (and dry stone walls) require the least amount of technical know-how. The work is easy because you can proceed at your own pace and there is no need for a crew. One person can easily work alone without any problem.

PLANNING

Determine, if you have not already done so, the primary purpose of your walk. From this decision, you can derive its minimum width and an acceptable construction material.

Width. Following is a list of minimum walk widths that have been found satisfactory. Note that many municipalities install sidewalks, so check this before you unwittingly donate a sidewalk to the city.

MINIMUM WALK WIDTHS

Public sidewalks	5 feet
Walk to front door	3 feet
Service walk	30 inches
Garden walk	2 feet

TYPES OF WALKS

For the purpose of discussion, we can divide walks into four types: stepping-stone, flagged, concrete and flagged concrete. A stepping-stone walk consists of stones or what-have-you spaced a distance apart so that one has to literally step from one stone to another. A flagged walk consists of flags—stones or

Fig. 68. Garden walk made of face brick laid up on sand. The walk is very old and the bricks have sunk into the earth until their surfaces are level with the grade.

any material that can be laid upon the ground and stepped on—placed side by side on the earth. Concrete walks need no explanation; they are simply long slabs of concrete poured into shallow forms laid on the earth.

Flagged concrete begins as a plain concrete slab, then covered with flags of various sorts permanently mortared in place. The reason for the flagging atop the concrete is merely visual: flagged concrete looks much better than bare concrete.

(36)

Fig. 69. An edging tool.

Fig. 71. A homemade wooden float.

Fig. 70. An aluminum float.

Fig. 72. A caulking trowel.

TOOLS AND EQUIPMENT

You'll need line, pegs and a steel tape or folding rule for laying out the walk. For flags on soil, you'll need the stone cutting tools mentioned previously. For brick or block walks, you will need the tools listed in Chapter 3. If you excavate, you will need a shovel or grub hoe to remove the sod, and a hammer, saw and spirit level to build the form, as well as a rake to level the soil.

For a concrete walk, you will need the digging and form-making tools plus the concrete-making equipment suggested in Chapter 2, as well as a screed, rake, float, edger and groover. For flagging the concrete, you will need whatever tools have been mentioned previously for cutting your flags, be they stone, brick or concrete, plus mixing tools for mortar.

STEPPING-STONE WALKS

We will start here because these are the easiest to make of all walks and can be completed in a few hours.

Materials. You may use 2-inch flagstones (1-inch stones will crack), 2-inch-thick wood blocks, or fieldstones with one reasonably flat side. All of these should be a minimum of 1 foot square, as smaller flags tend to turn underfoot. You may also use 2-inch or thicker solid concrete blocks.

Procedure. Cut the grass as close as possible to the ground along the desired path, and rake the area

clean. Lay the 2-inch flags directly on the ground. Space them no more than 18 inches apart, center to center. Arrange them in a straight line or however you wish. In a short time, each flag will sink a bit into the earth, locking it in place, and the grass will grow up between the flags, giving it all a finished, permanent look.

Fieldstones, because they are uneven, require more work than flagstones. Examine the underside of the fieldstone. Dig a shallow hole in the earth that conforms to the shape of this side. Try the stone in the hole: you want the top of the stone to be level and about 2 inches above the soil. The reason for the 2 inches is that this is the usual height of cut grass, and you want the top of the cut grass to be more or less flush with the tops of the stones.

FLAGGED WALKS

Flagged walks differ from stepping-stone walks in that the flags are either touching each other or very closely spaced. Depending on the thickness of the stones and the decision of the mason, the flags may rest on the surface of the earth, on the surface of excavated earth or on sand placed on the excavated surface.

Start with the flagging. Since the dimensions of the flagging will determine if and how much you may want to dig or need to dig, as well as the width of the walk, start by selecting your flagging. Lay the stones out side by side, spaced however you wish, and in a pattern or not, as you wish. Then measure to see what walk width works out best and does not require flags to be cut.

Basic rules. Flags larger than 1 foot square and thicker than 2 inches can be laid directly on the earth without fear of them shifting much or breaking. They can also be spaced a distance apart without problem. Flags 1 inch or less in thickness are best not laid on top of the earth; with these it is best to excavate a bit to provide a soft and flat support. Flags less than 1 foot square should be butted against each other so that they will act to hold each other in place. But no matter what the flag's area or thickness may be, the most level and stable walking surface will be secured by excavating and supporting the flags on a layer of sand.

Surface flagging. Cut the grass and sweep the path clean. Position the stones. Use two parallel, stretched mason's lines as guides to keep the flags in a straight line. Use a thin stick as a quick spacing guide. To make a curve, drive a peg in the ground and attach one end of a line to the peg. Then, with the other end of the line pulled taut against the peg, walk in an arc. Your path will give you the curve you need.

Excavated flagging. With the help of two parallel mason's lines and pegs, lay out the desired walk—plus 1 extra foot of width (six inches on each side). Excavate the area between the two lines with a straight-edged shovel to a little less than the necessary depth. Next, construct a wooden form (that you will use as a guide). Make the form of either 1 × 4 or 2 × 4 lumber and 1-foot-long pegs. Construct the form within the excavation by driving the pegs into the earth and nailing the lumber to the pegs. Make the form a fraction of an inch wider than the desired walk, with the pegs to the outside and the top edges of the form level and 2 inches higher than ground level. The top surface of the form will thus be the guide that establishes the top surface of your flagging.

Rake the surface of the soil loose and free of stones, and then lay the flagging in place. As you work, check the top surface of each flag against the top edge of the form. Do this by placing a flat board across the form: each flag should just clear the underside of the board. If it does not, correct by excavating a little more beneath it. When all the flags are in place, space between them can be filled with loose soil or sand. Finally, the form is removed and dirt and sod piled up against the sides of the flagging.

Flagging on sand. Proceed as before, but dig down another inch. Spread an inch of sand over the bottom of the excavation and level it as best you can. Lay the flagging on top of the sand, using the form

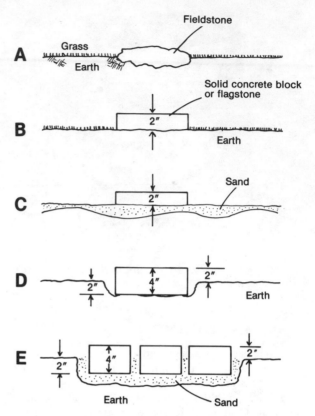

Fig. 73. *Excavation techniques for use with various flagging materials: A. Excavation to match underside of fieldstone. B. Where earth is fairly smooth, blocks are laid directly in place. C. Where earth is not smooth, hollows can be filled with sand. D. When thick flags are used, depth of excavation is altered to produce desired walk-surface elevation. E. For a smoother walk surface, excavation is partly filled with a layer of sand, which can be easily leveled.*

as above, and filling in the spaces with dirt in the same manner.

CONCRETE WALKS

Concrete walks differ from the others in that they are not flexible. For example, they cannot rise and fall with frost heave without cracking. Therefore, in wet areas concrete must be drained. Also, since concrete is watertight, such walks must be pitched either lengthwise or to one side. If not, water remaining on the surface can turn into dangerous ice.

All concrete walks should be at least 4 inches thick. Therefore, you will need to dig down a mini-

Fig. 74. Walk made by laying 2-inch-solid concrete blocks directly on close-cut grass. Lines hold the walk straight. Stick provides rapid and accurate spacing.

Fig. 75. Herringbone-pattern walk made with 2-inch-solid concrete blocks. Blocks will sink down a bit and the grass will soon fill spaces in between.

mum of 2 inches. If the soil is soft (black with humus) or you know that it is often water-soaked for periods of time, you should dig down another 4 inches to allow for a 4-inch-thick layer of gravel or crushed stone. The stones will help drain the earth beneath the concrete and help keep it from cracking in the winter. Woven wire mesh is also used, mistakenly, for the same purpose. The wire does little to eliminate cracking; what it does—and this is an important function—is to keep the cracks from opening and the pieces of concrete from spreading apart.

Excavation and form construction. Start by laying out the walk with parallel mason's lines 1 foot wider apart than the walk is going to be. Construct a form within the excavation similar to that described for making flagged walks, close the ends of it with additional boards. Remove or tamp down the high spots, or fill in the hollows with crushed stone or gravel (Fig. 77). If you must fill with soil, tamp firm; otherwise, the replaced soil will subside in time and you will have an unsupported area beneath the slab.

Tamping. This is the process of tapping (or even pounding) downward on the soil to compress it. Compression hardens the soil, making it less permeable and therefore less likely to absorb water, or as much water as when it is relatively soft. Tamping is

not nearly as effective as a layer of crushed stone beneath concrete as far as crack prevention is concerned, but it does help.

You can tamp with a rented metal tamper, or a homemade tamper consisting of short blocks of 2 × 4s nailed to the end of a long 2 × 4 that is used as a handle (Fig. 78). The smaller the bottom of the tamp, the more you will compress the soil, but the longer the job will take. There are also powered tampers that can be rented.

You will find the tamper very useful when you are bringing the surface of the excavated trench down a fraction of an inch. It is easier to pound the high spots down than to skim them off with a shovel. You will also find tamping will go much more rapidly if you first wet the soil down a little.

Next, place a straight-edged board across the width of the form, with a spirit level on top of it. If the walk doesn't run downhill, raise or lower either side of the form to secure a pitch of about ¼ inch per foot. See that there is one stake behind every 3 feet or so of form, and that the tops of the stakes do not project above the top edges of the form.

If the walk abuts another walk or any other solid masonry, you must insulate the two with an expansion strip. This is a black strip of fiber 4 inches wide

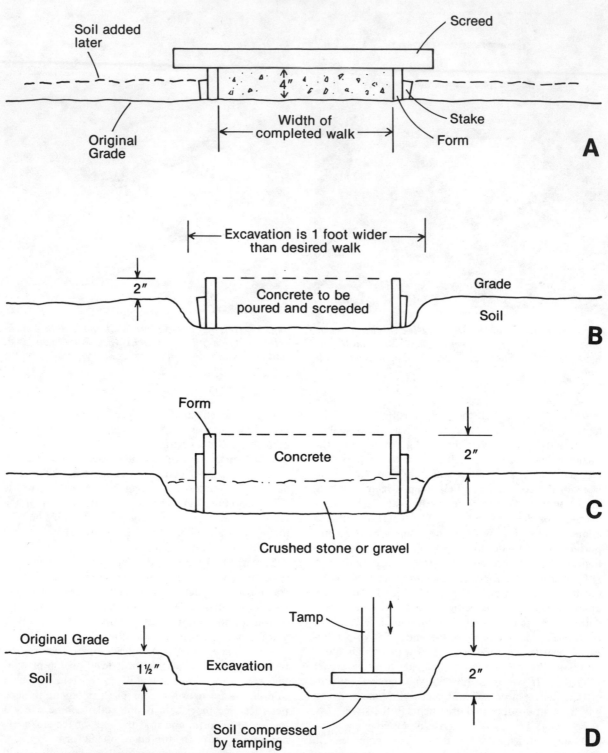

Fig. 76. Excavation techniques for use with concrete walks: A. When soil is firm, slab is poured directly on top. More soil is added alongside the slab later. B. When soil is soft, a trench is excavated for use with form. Depth should leave top of finished concrete 2 inches above grade. C. When soil is soft and wet, the trench is dug to a depth sufficient to accommodate crushed stone or gravel base. D. Trench is only partially excavated. The last inch or so of necessary depth is produced by tamping.

Fig. 77. Crushed stone has been spread evenly across the bottom of the excavation, and a ruler is used to make certain that desired concrete-slab thickness is maintained. To check pitch of slab from side to side, a spirit level may be placed atop the board shown.

Fig. 79. Using steel wire mesh to reinforce a concrete walk. Mesh is placed atop crushed stone; when the concrete is poured, mesh will be lifted a little so as to more or less center it in the concrete.

Fig. 78. A tamper is made by nailing short 2 × 4s to a longer 2 × 4 used as a handle between them.

Fig. 80. Curved form raised on stakes and filled to 4 inches below its top with crushed stone. Boards across form are stop boards, which create expansion joints. When the concrete has set, they will be removed and the resulting space filled with tar.

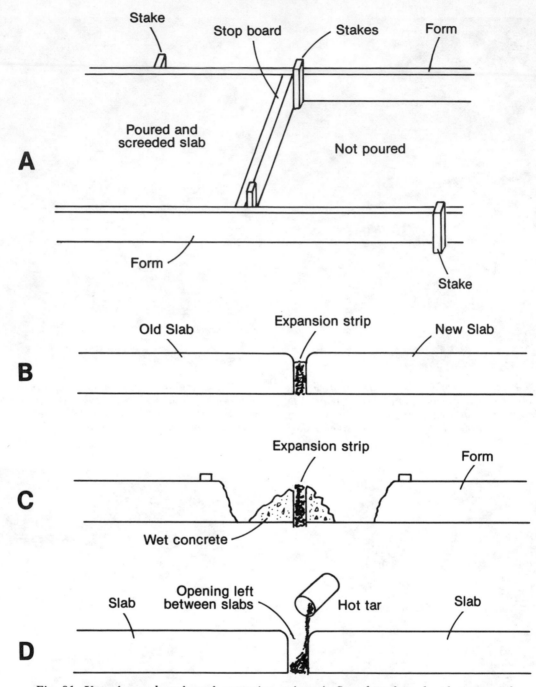

Fig. 81. Use of stop boards and expansion strips: A. Stop board is placed across end of slab perpendicular to form so that you can stop work at this point. B. When the slab you have poured has hardened, the stop board may be moved on and an expansion strip placed as shown so that the next slab you pour will be separated from the first. C. To place an expansion strip across the width of a form without the use of a stop board, position it, as shown, with some wet mortar, then pour mortar carefully on either side. D. Another method is to remove the stop board and fill the space with hot tar.

(42)

and several feet long. It is flexible, waterproof and rotproof, and you can easily cut it with a knife. If the walk is going to be 25 feet or longer, it should be divided into two shorter lengths by a transverse expansion strip. Whether insulating the end of the walk from another walk or similar structure, or dividing the walk into two, the strip is positioned transverse to the walk with its top edge ¼ inch below the top edges of the form. Use globs of mortar or concrete to hold it in place until you pour. If you want to pour up against the strip, place a board supported by stakes behind the strip.

Unless you are very strong, it is best to have a helper when you pour and screed any form wider than 3 feet. If you can get two helpers, the entire job will be speeded up several times over.

Mixing and pouring. Take one last look at everything to make certain the form is in its proper place, that it is properly pitched and braced and that you have all the necessary tools on hand. Use stones 1 inch and smaller and preferably the 1:2¼:3 mix. If you are working alone, mix no more than 2 cubic feet at one time. With a helper you can go to about 4 cubic feet; with three on the job you can probably handle 8 or 10 cubic feet at a time. If you are working with Ready Mix, do not tackle the job with less than three adults. It is much too difficult.

Start your pour at either end of the form. Use sufficient water to secure a soft mix, one that can't be piled more than a couple of inches high. Dump the concrete gently into the form. With a hoe, shovel or rake, push the concrete up against the end of the form and into the corners. Puddle the concrete a bit with your chosen tool so that it settles into place.

Screeding. Place the screed (any strong, straight-edged board that can reach across the form by more than a foot or two) on its edge across the form. Start at the end of the form which is now filled with concrete. Press the screed down on the form and slide it back and forth like a saw. At the same time, move it forward (Fig. 82). This is the basic screeding operation, and it makes the surface of the concrete level with the top of the form. If there is a hollow spot, fill it with concrete. Lift the screed, return it to its original position, and screed some more. Note that the screed is always brought against the concrete in the same direction.

Add more concrete and continue screeding. Do this until you have completely filled the form. If you have temporarily run out of concrete, just work the screed past the end of the concrete. You can let as much as twenty-four hours pass before continuing, if the weather is cool and moist. If the day is hot,

Fig. 82. The basic screeding operation: sawing the screed board across the form while forcing it forward brings the top of the poured concrete level with the top of the form.

windy and dry, you should not let more than a couple of hours pass; otherwise, there will be a visible demarcation between the new and old concrete. If days pass, you should apply Weld-Crete or a similar chemical bonding agent to the end of the old concrete. This will help the new concrete adhere to the old.

When you know you are going to halt the job for a day or more, the better procedure is to place a stop board across the form at that point. Make the board square with the sides of the form and its top flush with the top edge of the form. This produces a neat, square end to the slab at this point. When you resume pouring, remove the stop board and place a length of expansion strip against the old concrete, (Fig. 81), and then continue pouring, or you can just pour without the strip in place. In the latter case, edge the concrete at the stop board before removing it. (Edging is discussed in the following paragraphs.)

Finishing. Bear in mind that though you may not want to complete the job in one day or less, you must finish what you have poured within an hour (less on a hot, dry, windy day), or you will end up with a strange piece of horizontal sculpture. Finishing consists of tamping, edging, grooving and floating, which is a type of smoothing process.

For small areas such as walks, you can tamp with the end of an iron rake held in a vertical position and moved up and down against the surface of the concrete (Fig. 83). The purpose of this type of tamping is to push the stones down beneath the surface of the concrete. As long as they are out of sight, that is all that is necessary. If you have difficulty tamping you have waited too long or your mixture is too dry. (Too late to add water now. Just tamp harder or let it go, only appearance is affected.) If you have a large area to tamp, it is best to secure a commercial tamper called a Jitter Bug. Some masons forgo tamping and hope the stones will sink beneath the surface of the concrete and out of sight by themselves, which they sometimes do when a high-sand mix and a little extra water is used.

If tamping has roughened the surface of the wet concrete, screed it again, then wait for initial set. This is when excess water within the slab comes to the surface. At this time you float the surface of the slab. A float is a rectangular trowel made of wood or aluminum, applied with a circular pressure. Its leading edge is held slightly raised so that it does not dig into the concrete (Fig. 84). If your efforts produce no visible results, the mix is still too wet; wait until the concrete feels like wet sand beneath the trowel. If your efforts cannot remove the lumps and bumps, you have waited too long (sometimes wetting the concrete down helps a little). In any case, the entire surface must be floated.

Next, the slab (or whatever portion has been poured) must be edged. This is done with an edger. The blade of the tool is forced between the slab and the form and slid along with a little downward pressure (Fig. 85).

Following, the slab is grooved. This is done with a groover and a board guide. The board is placed across the slab at right angles to the form, and the groover is pressed into the concrete and slid across (Fig. 86). Doing so leaves a groove, which acts as a crack controller. If the slab is going to crack, it will most likely crack at the groove. Generally, the grooves are spaced so as to divide the walk into squares. If you don't have a groover, you can make do with an edger, by sliding it in two directions across the slab. If necessary, you can give the slab another little touch of the float here and there.

Curing. Curing follows grooving with as little time lost as possible. It consists of covering the green slab with wet newspapers, wet straw, wet sand or a sheet of plastic to reduce moisture loss by the concrete. As an alternative, you can keep the slab wet with a con-

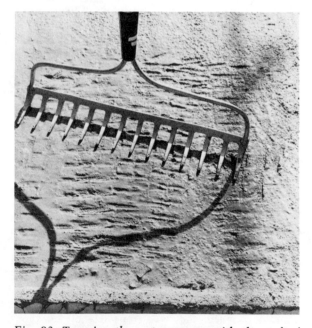

Fig. 83. Tamping the wet concrete with the end of an iron rake. This drives the stones beneath the surface.

Fig. 84. Using an aluminum float.

Fig. 85. Using an edger.

Fig. 87. A curved slab that was produced by the curved form, now fully cured.

Fig. 86. Grooving a slab. The tool is moved across slab in two directions, using the board as a guide.

stant fine spray of water. Generally, the moisture barrier is kept in place for two or three days, but a week would be much better. Following curing, the barrier and the form can be removed and the walk can be used.

If you want to do your job piecemeal and still have it look good, place a stop board across the form wherever you wish to stop, then finish that section of walk. In this way you will have a neat job without having to complete the entire walk.

FLAGGED CONCRETE

A flagged concrete walk begins as an ordinary concrete walk, and is then covered with a layer of flagging set in mortar. Any kind of flagging except wood can be used. Usually the flags are smaller than the width of the walk, which permits them to be positioned in patterns.

Excavation depth. Begin by deciding on the type of flagging. Determine its thickness. Add 1 inch to this figure, then add another 2 inches. This will give you the required excavation depth necessary to bring the top of the finished walk 2 inches above grade, which is the surface of the soil by its architectural name. Assume, for example, that you wish to cover the walk with 1-inch flagstone. Allowing 1

(45)

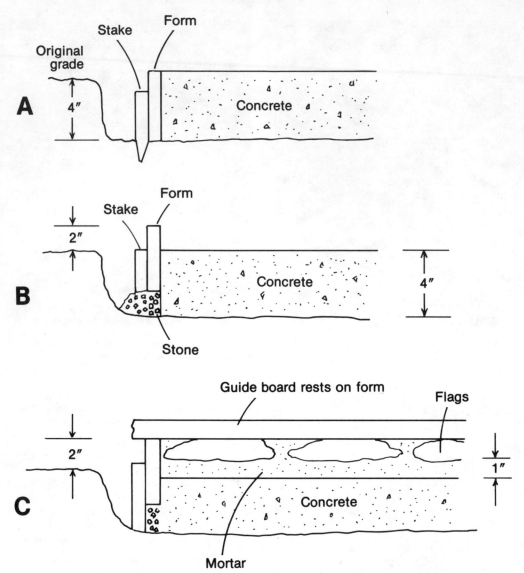

Fig. 88. Steps in constructing flagged concrete: A. The trench for the slab is dug to a depth that will make the top of the slab flush with the original grade. The form is then constructed, and the concrete poured and screeded. B. When the concrete has hardened, the form is raised 2 inches. Small stones are used to keep the form from settling. C. The concrete is covered with a 1-inch layer of mortar, and flags are positioned atop the mortar. A board placed across the form serves as a height guide. High flags are pushed down; low flags are raised by placing additional mortar beneath them.

inch for the flagging, 1 inch for the mortar and 4 inches for the concrete makes for a total of 6 inches. If you place this combination into a 4-inch trench, 2 inches will protrude, which is what you want. If you were to use 2-inch flags, the figure would be 2 + 1 + 4 to produce 7. A 5-inch trench would be just right for this combination.

Construct the slab. Proceed as if you were going to make an ordinary concrete walk, but dig as deeply as necessary to accommodate the flagging and

Fig. 89. Bottom of flag had been sprinkled with water and then lowered into position, but it proved to be too high. Now some of the "dry" mortar is being removed before trying the flag in its position again.

Fig. 90. This flag is only a little too high. Pounding it down with a small sledgehammer and a piece of timber may be all that is needed.

the supporting mortar. Pour, screed and tamp the slab as previously directed; float, but don't trouble to edge it. Let the slab cure for two or three days, then carefully raise the form 1 inch for the mortar, plus the thickness of the flags you are going to use. With stones and/or stakes, lock the form in place so that it cannot sink back into the earth. The top edges of the form are now your guide. When the flagging is in place, the surface of the flags should be exactly level with the top edges of the form. There are other ways of doing this, but this method is simplest.

Prepare the mortar. For flagstone and other non-porous material, a one-part cement, three-part sand mixture is used. For brick and porous materials, mortar cement is used (cement with lime added). Mix the sand and cement very thoroughly, then add just enough water to permit you to form the mortar into an almost dry ball that will hold its shape. When the mortar is pressed, a little water may be exuded, but no more.

Laying the flags. Begin at the end of the walk, and wet 2 or 3 square feet of the concrete with water applied with a brush. Cover the moistened area with a layer of mortar 1 inch thick. With a small stick or similar tool, make the surface of the mortar as level as possible. Wet the bottom of one flag with water

and sprinkle a little water on the mortar that will support the flag to be laid. Lay the flag in position. Tap it down lightly with the heel of your hand. Place a straight-edged board across the form and just above the flag. If the flag is properly positioned, its top surface should just miss the underside of the board. If it does not, remove or add mortar as necessary. Rewet the underside of the flag and try it again. If the flag is just a fraction high, you can sometimes force it down by placing a short 2 × 4 on top and striking the wood with a small sledgehammer.

Continue this way down the length of the walk, taking care not to spread so much mortar that it dries before you get to it, and that you position the flags in the pattern you wish.

Finishing. Give the flags three or four days before you try walking on them; otherwise, you may break them loose. Then fill the joints, using the "dry" mortar mix suggested previously. With a brick trowel or a caulking trowel (which is long and narrow-bladed), press the mortar firmly into the joints; this will compact the mortar and give it a smooth surface (Fig. 91). Wait a few days for the mortar to dry before walking on the flags.

Finally, remove the form and check the sides of

Fig. 91. Dry mortar has been placed in the joints between the in-place flags. The point of the brick trowel is used to compress the mortar and make it smooth.

the walk. There may be open areas between the flags and the concrete. Force mortar into these holes. Make the edges of the walk smooth with the trowel. Fill the space between the unexcavated earth and the walk with soil. Seed this area if you want to speed the growth of the grass.

CHAPTER SEVEN

Driveways

Driveways differ from walks in that they are of a necessity much larger and in that they are generally regulated by the local municipality by means of the local building code. The very size of even a single-car driveway is such that it is almost impossible for a single individual to do the job. At a minimum, two people are necessary, and to make the work reasonably easy you need at least three.

BUILDING CODE

Most municipalities limit driveway slope or pitch to 15%, and some to as little as 10%. Most municipalities also regulate the profile of the driveway as it passes a sidewalk (or the site of a future sidewalk) on its way to connect with the road.

Slope. This is the angle of a road, or whatever in relation to the horizon. Sometimes this is expressed as drop per foot, but most often it is expressed as a percent, which means drop per 10 feet or 100 feet. The higher the figure, the steeper the slope.

Measuring slope. You cannot estimate slope. Your eye will deceive you. It must be measured. This is easily done (Fig. 92).

To measure slope, drive a long stake vertically into the ground (or have a friend hold it for you) at the bottom of the grade. Measure from where the stake enters the earth to a point 10 feet up the hill. Mark this point. Stretch a line from this mark to the vertical stake. Hang a line level on the line. Raise or lower the stake end of the line until the bubble is centered. Mark the position of the line on the stake. Measure down the stake to the earth from this mark: this is the drop or slope per 10 feet. If the drop is 1 foot, you have a 10% slope. If the drop is 1½ feet, you have a 15% slope, and so forth. Dividing the drop by the length will always give you the slope.

Slope correction. If the slope of your drive is too steep at one or more places, you may be able to remove some soil and convert the short, steep slope into a longer, gentler slope. If you have not as yet begun construction of the garage, you might consider lowering the garage. This is better than not getting a building permit. Another possibility is to extend the drive, perhaps adding a sweeping curve so that the drive does not follow the shortest and steepest route to the road. Still another possibility is to ask the city for a variance.

PROFILE

Many municipalities have codes regulating the profile of the drive where it approaches and passes the sidewalk. The drive's profile literally means a side view of the drive at that point (Fig. 93). Obviously, if there is a sidewalk, you must construct your drive so that its top surface is flush with the walk. If there is no walk, you will have to learn where the walk's surface eventually will be, so that when the city does install the walk (this is almost always the province of the city), your drive will not be too low

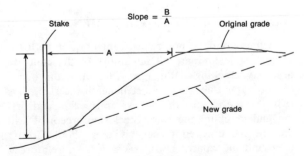

Fig. 92. *How to measure slope, and one way to reduce excessive slope.*

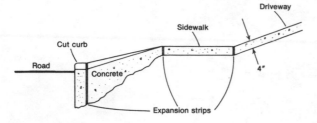

Fig. 93. Profile of a driveway as it crosses a walk. Note the expansion strips that isolate the drive from the walk and the curb.

or too high at that point. If there is a curb, you have to secure permission to cut it, and when you do, you must cut it to city specifications. Naturally, if there is no curb, you must anticipate it correctly.

DIMENSIONS

Drive width. A single-car drive should be at least 10 feet wide. A two-car drive should be twice that. When a single-car drive has to widen to meet a two-car garage, the drive should begin to widen 16 or more feet from the garage. If the width of the garage door or doors is greater than the minimum recommended drive widths given, increase the drive widths so that there is no chance of a car backing out of the garage and driving off the concrete.

Thickness. Unless you are expecting trucks, there is no need for concrete more than a nominal 4 inches thick. If you believe that trucks and passenger cars may back into your driveway for the purpose of turning around, and if the last portion of the drive is steep (making cars bounce a little at this point), it is wise to increase the thickness of the concrete to 6 or more inches for the drive's last 2 or 3 feet. If the driveway is to accommodate trucks, it's best to make the concrete 6 inches thick for its entire length.

Height. The surface of the drive should be about 1 inch below the entrance to the garage. This is done to reduce the possibility of water flowing into the garage. At its other end, the drive should be about an inch higher than the cut curb.

Pitch. When the drive runs downhill for 20 feet or so, the drive does not need a side pitch. When the drive is much longer than this, it is best to pitch the drive. Doing so dumps some of the rainwater off to the side of the drive and reduces the flood of water running downhill. On perfectly flat ground, pitch is a must or you will have sheets of water on your driveway, and ice in the winter.

On a 10-foot-wide drive, you can secure a "fast-flow" ¼-inch-per-foot pitch by simply lifting one side 2½ inches higher than the other. More is unnecessary, and a little less will do. Since the low side of the drive will be 2 inches above grade (so that the cut surface of the grass will be flush with the surface of the concrete), the high side will be 4½ inches above grade. This is obviously too much. Correct by placing soil alongside the high side of the drive and sloping it gently down to the existing grade.

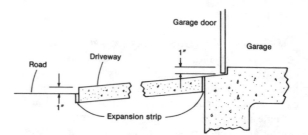

Fig. 94. Driveway end details. When the drive does not contact a curb, it should be about 1 inch higher than the road and isolated by an expansion strip. The garage end of the drive should also be isolated and about 1 inch lower than the garage floor.

On a 20-foot-wide drive, a pitch of ¼ inch to the foot brings the high side 7 inches above grade. This is too much to hide without leaving a noticeable slope. The solution is to split the drive down its length with an expansion strip. The strip helps control cracking, and by dividing the slab in two greatly eases construction, since it is much easier to pour and screed a 10-foot slab than a 20-foot slab. To secure the desired pitch, the center of the drive is raised the necessary 2½ inches.

DRAINAGE AND REINFORCEMENT

Neither drainage nor reinforcement (nor both together) will eliminate all possibility of the concrete driveway cracking. Drainage will reduce the quantity of water that may collect beneath the slab, thus reducing the chance of said water freezing and lifting and cracking the slab. Steel reinforcement will prevent the cracked pieces (if they should form) from moving apart. Thus, a crack in a reinforced slab does not necessarily result in the eventual destruction of the slab.

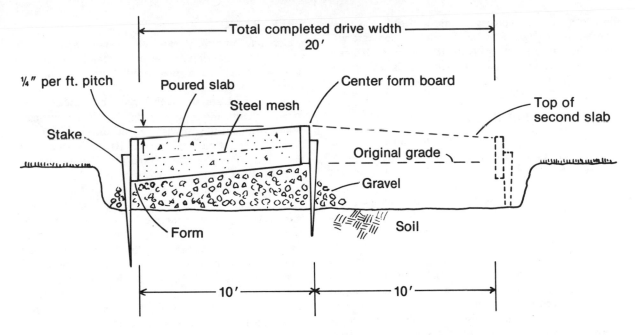

Fig. 95. *Cross-section of a double-width drive under construction. When the poured slab has hardened, the center form board is replaced with an expansion strip. The other side form is then installed and the second slab poured, using the top of the hardened slab as a reference.*

Naturally, both drainage and reinforcement increases the labor and cost of constructing a driveway. When these additions in effort and expense are warranted is difficult to ascertain, and the best we can do is provide some general guidelines:

First, examine the terrain to determine water runoff. Also, determine frost conditions and examine the soil after removing the sod (grass). If the drive is to be positioned on low-lying ground, or on a shallow depression where rainwater collects and remains for a day or so, it is best to lay down a granular base (stone) beneath your drive and to reinforce it with steel. If the same location is subject to frost, there is no sense in constructing a concrete drive without both stone and steel. To ascertain frost conditions, contact your building department or ask your neighbors.

If the soil is dark, it is soft and filled with humus. On such soil it is necessary that at least a stone base be used, with or without the presence of frost and standing water. If you *expect* frost and water, be certain to use both stone and steel.

If the soil is light in color, it is hard. Neither stone nor steel is needed here (frost or no frost), so long as you do not expect standing water. If there is sand

and you expect standing water it is best to use both stone and steel.

Granular base. As stated, a granular base provides drainage beneath the concrete slab. You can use gravel, crushed stones or cinders. Their sizes are unimportant; however, you will find it difficult to work with stones much larger than 1 inch. The base is simply spread over the bottom of the excavation and made as level as is practical.

Steel reinforcement. Concrete drives and walks are reinforced with welded steel wire mesh. Size 6 × 6-10/10 is the size most often used. It is sold in rolls 5 and 6 feet wide. It can be cut with a bolt cutter or a hacksaw, but cut or uncut, it is a forest of sharp metal edges. Always wear leather-palmed gloves when handling this stuff.

The wire mesh is positioned within the slab, its edges always 4 or so inches short of the edges of the slab. To install, the mesh is unrolled and laid within the form. Since it tends to retain its curl, you can stomp on it to flatten it out. To position the mesh within the slab, you can either lift it up a couple of inches with the help of some small stones and then pour over it. Or you can pour first and then lift the mesh up a little to permit the concrete to flow un-

(51)

derneath. To join pieces of mesh together, simply overlap their edges 6 inches or so. The concrete will lock the edges together. Note that wire mesh never goes beneath an expansion strip, but is always cut short of it by 4 inches or so.

TOOLS AND EQUIPMENT

You will need line, pegs and a steel tape for laying out the driveway, and a folding rule may also be helpful. A shovel and grub hoe can be used for removing the soil, grass and roots, and a wheelbarrow for moving it all away. A saw, hammer, line level and spirit level are needed for making the form, and a razor knife for cutting the expansion strip.

Concrete mixing and placing equipment, as discussed in Chapter 2 is essential, plus a bull float or darby, a wood float, and a long, straight-edged board for screeding. If you have to cut the curb, you will need a mason's chisel and a small sledgehammer. Also, don't forget overshoes for yourself and your crew.

Fig. 96. The line level. With care, you can secure accuracies of better than 1 inch in 50 feet.

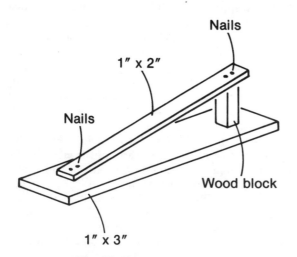

Fig. 97. Homemade darby.

PREPARATION

Concrete driveways are constructed very much like walks. The differences arise mainly from size and the problems that may result from adherence to building codes, if any.

Since drives are so much larger than walks, a lot more concrete must be mixed, placed and finished. While you can, if you wish, divide the slab into sections; each section will still be 10 by 10 feet at a minimum. It is therefore foolish and dangerous to your health to attempt to construct an entire concrete drive by yourself. Excavating and form construction can be done by yourself at any pace you choose, but once the concrete has been mixed, it has to be placed, screeded, tamped and finished in an hour or so. To attempt to do this with less than three people is looking for a broken back or worse. With three it is easy, even for the inexperienced.

And since drives are not narrow, you cannot reach across them from any point. For easy working, it is advisable that the entire crew be prepared with overshoes or boots—the kind without buckles and that reach the knees are best. With overshoes, you can all muck about in the concrete as necessary without straining your backs reaching out. Hose the overshoes down before the concrete has dried, and they will wash clean.

EXCAVATING

Begin by outlining your projected driveway with two parallel lines spaced a foot or more farther apart than the desired drive width. For a 4-inch-thick slab, go down 2 inches; for a 6-inch slab, go down 4 inches. If you are going to lay down a stone base, go down another 4 inches. To be of any value, the base should be approximately this thick, but in any case you have to dig deeper than the grass and its roots. You should never lay a drive on grass or its roots.

If you have accidentally dug too deeply at any spot, or you were forced to dig more deeply than you needed because of roots, you cannot fill the hole with soil (unless you tamp it down), as it will be too soft. You must fill the depressed area with either concrete or stone (any washed-clean stones will do).

Now is the time to cut the curb, if necessary (Fig. 98). Secure permission and instructions from the building department, since they may specify the depth of curb cut. Then cut away with a mason's chisel and small sledgehammer if you have time and need the exercise. Otherwise, rent a jackhammer and a compressor to drive it.

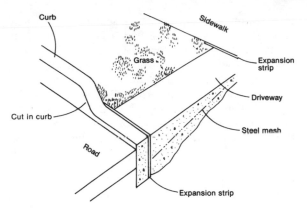

Fig. 98. Curb and slab-end treatment when drive terminates at a curb.

FORM CONSTRUCTION

Use 1 × 4 or 2 × 4 lumber for the sides of the form. Position heavy stakes at the outside of the form every 3 feet or so. Make or cut the tops of the stakes flush with the top edge of the form. When making a 20-foot-wide driveway in two lengthwise slabs, construct and install one side of one form first, then construct and install the second side 10 feet and 1 inch from the first. The 1 inch is where the expansion strip will be positioned. As mentioned before, make the top surface of the second (middle) side of the form 2½ inches higher than the top edge of the outside edge of the form. This provides the desired high center for the completed drive.

To check the comparative elevations of the form's sides when they are farther apart than the spirit level can reach, place a long board on edge across the form, and then place the level atop the board.

To make certain the sides of the form are straight along its length, stretch a line alongside the form. Since the stretched line is perfectly straight, you can easily see whether the form wanders from the straight and narrow. To make certain the top of the form is also straight—doesn't have hills and valleys—stretch a line above the form and check the clearance between the two.

Placing the stone. When possible, have the truck that is delivering the crushed stone or gravel back right into the form and then spread the stone by dropping it as the truck moves forward and out of the form. When making a two-car driveway in two halves, only one of the two forms need be erected at this point. So spread the needed stone in the completed form. And then, either have the truck drop the balance of the stone where it will be needed when the second form is completed, or have the truck make a second delivery.

Spread the granular stone base with a rake and shovel. Make it as level as practical and see that the stone extends beneath the form (and beyond, if you can). The top surface of the stone base should, of course, be flush with the bottom surface of the form. To make a 4-inch-thick slab, let your form boards be a nominal 4 inches high; for a 6-inch slab, you will need 6-inch-high boards.

Wire mesh. If you are going to use wire mesh reinforcement, it is laid atop the stone base. No allowance is made for its thickness. And, as previously stated, either lift the wire an inch or so with a number of small stones, or raise it a bit after it has been covered with concrete. Remember, the wire should be cut some 4 inches shy of all sides of the form, and should not be placed beneath any expansion strips.

Installing expansion strips. You will want a strip between the slab and the garage floor, a strip between the slab and the curb, and a strip between slabs when making a 20-foot-wide drive. When and where the strip is not high enough to reach from the soil or the granular base to the top of the form or curb, position a second strip edge on edge atop the first. You may have to cut one strip lengthwise to do so. Use a razor knife.

Placing the concrete. Use the 1:2¼:3 mix described in Chapter 2. If you are hand- or machine-mixing, mix no more than you and your crew can handle in an hour, including finishing. Use a wet mix that will almost find its own level. If you are going to work with Ready Mix, order no more than 3 yards if you are making a 4-inch slab or 4 yards when making a 6-inch slab. Both of these quantities will lay down 200 square feet. Thus, you will place and finish a 20-foot-long piece of concrete in one session.

To hold your slab to this dimension (or any other dimension), traverse the form with a stop board braced by strong stakes. In other words, construct a temporary end to the form at that point. Insert an expansion strip at that point later.

As you read this, a 20-foot slab of concrete 10 feet wide probably doesn't seem like much, and you may be tempted to save the extra cost of two small-load deliveries and opt for one instead. But on a hot, dry, windy day, when the concrete threatens to solidify into a stone sea replete with waves, you will be sorry you ordered more concrete than you could handle.

If your drive has a steep end-to-end pitch, it is best to begin concrete placement at the low end of the form. In this way, the concrete tends to remain in place better than if you start at the high end. In addition, it is also advisable to run the mix a bit on the dry side. This will also help keep the concrete from running out of the form.

Screeding. Select a straight-edged 2 × 6 at least 2 feet longer than the width of the form it has to span for your screed. Put a man or woman on each end of the screed, with yourself more or less at its center with a shovel. Then as they work the screed toward you, see to it that all the low spots are filled up with concrete, and that the concrete is puddled and forced into the corners of the form. When a hollow is filled, be sure to bring the screed back behind the filled spot and screed forward again. There is no harm in screeding more than necessary. But a missed high or low spot can cause plenty of trouble later on.

Finishing. Tamp the concrete immediately after you have screeded a small area. Just drive the top stones into the mud. You needn't do more.

Let the concrete set up, then float it with either a darby (a long wooden trowel) or a bull float (a giant trowel fastened to the end of a long handle). The best procedure is to begin floating with the bull float; since it is large it has the greatest leveling ac-

tion. Simply push and pull the float across the concrete. Its own weight is all that is necessary for it to do its job. Just make a few passes with the bull float, then go to work with the darby. Work with a long swinging motion, keeping the leading edge of the float up a bit to prevent it from digging in. Finish up with the hand float, or quit anytime after using the bull float if you are satisfied with the appearance of the job. If you wish, you can edge the slab exactly as you would a walk, although drives are not usually grooved, though there is no reason you cannot do so if you wish to.

As an alternative to a smooth finish, you can striate the surface of the concrete with a street broom, which is a coarse-bristled brush. This is sometimes called a broom finish. It provides extra traction on hills, as well as a surface with a different appearance.

Start by floating the concrete as described, then before any time passes, drag the broom across the concrete, taking care to keep the passes parallel. A single pass in a single direction is all that is needed. If you want deeper grooves, drag the broom across the soft concrete a second time in the same direction.

Curing. Cure as previously described. Cover the slab with anything that will either keep it moist or hold the moisture in. The simplest means is a large

Fig. 99. Using a homemade bull float. Its very size makes leveling easy.

Fig. 100. Running a broom across poured and floated concrete striates its surface. This reduces the possibility of cars and pedestrians slipping.

plastic sheet, which is sold at mason supply yards. Give the slab at least a full week to harden before permitting vehicular traffic. Two weeks is even safer.

The second slab. If you are constructing a two-car drive and have been following instructions, you have just completed one half of your drive. What you have now should reach from road to garage and should be just 1 inch less than half the full width of the completed driveway. The next step is to remove the entire form. Following, rebuild the form next to the finished slab. Do not, however, install two sides. Instead, the in-place, finished slab is used as one side of the second form. Before you pour the second slab, position an expansion strip up against the side of the completed slab. Make the top of the strip flush with or slightly below the surface of the concrete. Make the strip extend the entire length of the slab. To do so, butt the end of one strip against another. Again, use some small stones or gobs of concrete to hold the expansion strips in place.

Spread the granular stone base and lay down the wire mesh (if you are using them). Pour and screed, using the top of the now hard and tough concrete slab as one guide for the screed, and the form side as the other guide. Finish as before.

GRID PAVING

If you don't mind grass growing up through holes in your driveway, or if you would like to have grass growing where you are now thinking about installing a slab drive, you can build your drive out of grid-type concrete paving blocks. These are pierced blocks in various designs that are laid on the surface of the earth. Some soil may be spread over them and into the holes. Eventually, grass will grow up through the holes and lock the blocks in place. Thus, the blocks can be used on sloping roads, and on hillsides to control erosion.

Installation. For best results (meaning the least chance of one or more blocks cracking when an auto passes over), remove the sod and lay the blocks down on soil that has been raked smooth and free of rocks. Drop each block into place and then raise it and inspect the soil. The impression of the entire block must be visible; if a portion of the block is not supported, it may break under the load. Add and tamp down soil or remove soil as necessary. For big cars and trucks, it is best to lay down a 4-inch-thick layer of sand first and then position the grid blocks.

Fig. 101. Grid paving is easy to lay down and keep clean (just run a lawn mower over its surface). Courtesy National Concrete Masonry Association.

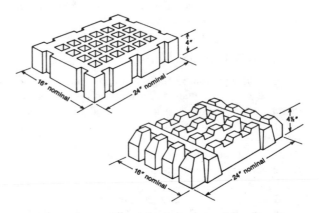

Fig. 102. Two types of grid paving: perforated and ribbed.

With the blocks in place, fill the openings with sufficient soil to make the surface of the soil within the holes level with the external grade. And that is it.

CHAPTER EIGHT

Patios

When you get down to basic design and construction, there are two types of patios. Both are simply paved areas near or adjoining a home or swimming pool. One consists of any type of flagging laid directly on the earth or a bed of sand. The other consists of a slab of concrete which may or may not be flagged with another paving material. Since the concrete-slab patio is by far the most popular, that will be discussed first.

Concrete patios may be of any shape and have openings for trees and other plantings. The patio's surface may be smooth concrete, painted, or flagged with any suitable material. In general, construction is similar to that of any concrete flat work—walks, driveways, floors, etc. However, local conditions and some designs may introduce important construction differences. It is these differences that will be emphasized in the following sections.

FLAGGED PATIOS

Like flagged walks, flagged patios may be constructed by laying flags directly on cropped grass, by covering the grass with a layer of sand and then laying the flags, and by excavating, replacing the soil with sand and then laying the flags atop the sand. The second and third methods result in respectively smoother and more stable patios.

Flags on grass. Nothing could be simpler. Cut the grass as close as you can, remove the clippings and debris and lay the flags down. Use 2-inch or thicker stones at least 1 foot square. Thinner stones will crack; smaller stones will turn underfoot (Fig. 104).

Flags on sand over grass. Cut and clean as before, and spread moist sand over the area. Drag a large flat board across the surface of the sand to make it smooth and level. You need only enough sand to fill the hollows in the earth.

Flags on sand in an excavated area. Dig sufficiently to remove the grass and its roots. Spread sufficient sand over the exposed earth to bring the tops of your flags to the above-grade height you desire. Level the sand and position the flags. Small flags 1 inch thick may be used here because the sand provides a solid, stable support.

On the whole, you can secure better results more easily if you construct a form encompassing your flagging area. Make the top of the form level and at the exact desired height of the flagging (use the form as a guide). Fill the spaces between the flags with soil and grass seed.

Fig. 103. Oversize face brick laid up in sand.

Fig. 104. A very simple and quick patio made by laying 2-inch-thick flagstone directly on cut grass.

Fig. 106. The sand has been made level and the flags positioned. A line stretched across the form helps the mason align the flags.

Fig. 105. When the grade is uneven, or when you want the flags to hold their positions for a longer period of time, the surface of the cut grass may be covered with sand (an even longer-lasting base can be secured by excavating and filling the excavation with sand).

Fig. 107. Sand is removed or added as necessary in order to correct the height of each flag. The surface of the flags is flush with the surface of the form.

CONCRETE PATIOS

It is a lot more difficult to remove a concrete patio than it is to construct it. It is also double the work to increase the size of a patio once it has been completed. And whereas walks and driveways are not places where one normally lingers, you will want to spend long summer days on the patio you build. Therefore it well warrants your time to give considerable thought to patio location, size and shape before doing anything else.

Location. Begin by listing your contemplated patio activities in order of importance. Do you want to sun bathe or dine there? For sunning, you'd naturally select the most sun-exposed area; for dining, shade would be preferred. Up north, the best location might be at the southern side of the house in order to extend the time during which you would be comfortable outdoors. In southern climes, the opposite would be true. Do you want easy access to the kitchen, or do you plan to do most of the cooking outdoors? All of these questions deserve careful consideration.

Size. A patio smaller than 10 feet by 10 feet is almost useless. This is the minimum area that will comfortably hold a table and six people. At the other extreme, a solid concrete back yard is certainly not a thing of beauty. The best approach to finding the happy medium is to lay out the patio with pegs and string and then place the furniture within the outline. Make believe you are using the space; sit, stand and walk around as if you and your family were actually using the patio. Then, if necessary, add a few feet so there is no need for anyone to step on the grass.

Shape. A single square or rectangular shape is easier to construct than a combination of shapes or a curved shape. At the same time, a curved shape is not that much more difficult to construct, and you may save a little concrete by the elimination of corners. Flagging a curved patio, however, is another tale entirely, as it will cost you considerably more time, labor and material than an equal-area rectangular patio. Flagged patios are discussed shortly.

TOOLS AND EQUIPMENT

You will need the same tools and equipment suggested for making driveways in the previous chapter if you merely lay down concrete. If you flag the concrete, you will also need the tools necessary for cutting and laying down flags on mortar. These tools are listed in Chapter 6.

If you are making a fairly large patio and plan to excavate more than a few inches, it may be advisable to rent a small bulldozer for a day. It is difficult to make a large surface flat and level with a shovel.

PRIMARY STEPS

Layout. Having determined the size and shape of your patio-to-be, lay it out with stakes and line, making the enclosed area 1 foot longer and wider than the desired slab. To make an arc or circle, drive a stake into the earth. Tie a line to the stake. Hold the line taut and walk it around the stake, marking your progress with pegs as you go.

Determine pitch. To provide fast runoff, you want a pitch of about ¼ inch to the foot in one or two directions across your patio. The actual pitch will be produced by adjusting the sides of the form accordingly, but at this point you need to know what pitch or slope, if any, exists. Do this by measuring up from the earth a distance of 3 inches on each stake, and marking the measurement. Next, stretch a line across the enclosed area from stake mark to stake mark. Hang a line level in the center of the taut line. If necessary, raise or lower one end of the taut line until the level's bubble is centered. Mark and measure the difference, if any, on the stake. Do this in both directions. You now know the pitch of the land, if there is any, and in what direction it may pitch.

Excavation. At the very least, dig deeply enough to remove all the grass and roots that may be present. If the exposed soil is soft and dark, if the soil becomes water-soaked at some time during the year, or if frost is a problem too, it is best to lay down a granular base for drainage. Since this base will be 4 inches thick, you will have to dig another 4 inches. Should you be planning to flag the concrete (this is discussed shortly), you also have to take into account the addition of the flagging and the supporting mortar. In any case, when you complete your digging you should have a flat-bottomed hole a foot or so larger all around than the slab you are going to pour.

CONSTRUCTING THE FORM

Build the form from 1 × 4 or 2 × 4 lumber nailed to stout stakes. Position the stakes to the outside of the form, with about one stake to every 3 lineal feet of form. Position or cut the stakes so that their tops are flush with or below the top of the form.

If the slab is to abut a walk, a stone wall or a

Fig. 108. Here two masons are using the form to help them screed the surface of the sand. Later, the form will be removed and the remaining spaces filled with more sand. Note that the sand has been lightly moistened.

Fig. 109. Interlocking concrete paving units are now laid atop the leveled sand.

Fig. 110. An example of what can be accomplished with concrete paving units laid up on sand.

building foundation, the slab must be isolated from said masonry by an expansion strip. The strip goes against the in-place masonry, and the concrete is then poured against the strip. The top of the strip is positioned flush with the top of the form.

To make certain a rectangular or square form has square corners, measure diagonally from corner to corner. When both diagonals are the same length, the corners are perfectly square.

Now check the pitch. Do this by either laying a straight-edged board on edge across the form and a spirit level on top of the board, or by stretching your line and line level across the form at several places in both directions. If the form is a bit high, drive it down. If low, pull it up and place stones underneath so that it will not sink down.

Curved forms. These are made from 4-inch-wide strips of ¼-inch plywood bent to the desired shape.

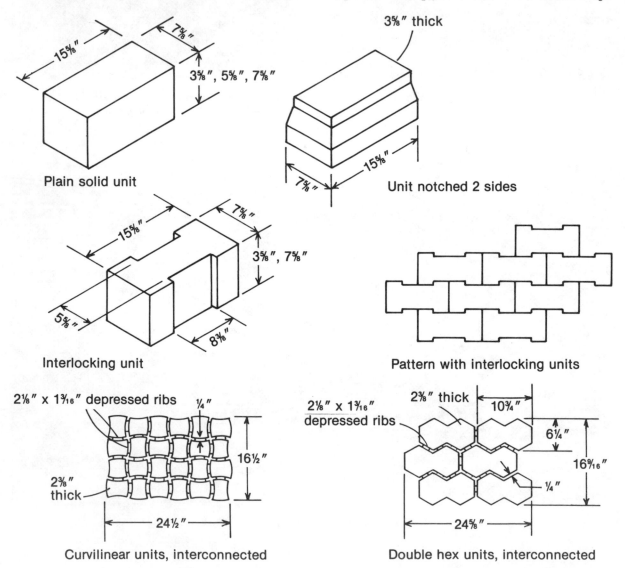

Fig. 111. A few of the concrete paving units available.

If the board will not bend as desired, score it—make a number of parallel saw cuts partially through the board—and then bend it. Since ¼-inch plywood is flexible, use plenty of stakes so that the concrete will not force it out of shape. To check the pitch of a curved form, use either method suggested above. Be certain to do this at several places in each direction.

Openings. These are open spaces entirely within the patio, made by erecting *enclosing forms* which exclude the poured concrete from the enclosed openings (Fig. 112). Enclosing forms are made from the

have to support the screed just as the outer form will.

Large patios. Patios that cover more than several hundred square feet usually introduce two problems not found in smaller patios. One is pitch: the wider the slab, the greater the difference in edge heights. With a pitch of ¼ inch to the foot, the difference at 10 feet is 2½ inches; at 20 feet, 5 inches; at 30 feet, 7½ inches. If your land is already pitched to some degree, much of the desired pitch may be there already; if the land is perfectly level, one solution is to

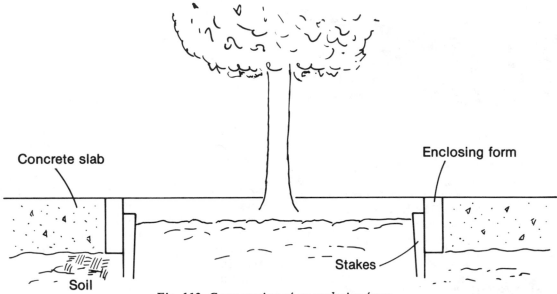

Fig. 112. Cross-section of an enclosing form.

same lumber used for the other forms, and supported by stakes the same way; however, the stakes go on the *inside* of the form. Form shape can be rectangular, square or curved. When positioned, the top edges of the enclosing form must be perfectly flush with the top edges of the outer form. To insure this, stretch a line across the entire outer form, positioned so that it also crosses over the enclosing form. The line must just clear the surface of the enclosing form. Adjust the enclosing form as necessary. Then, move the line and reposition it in the other direction across both the outer form and the enclosing form. Adjust the enclosing form again, if necessary. When properly positioned, the enclosing form's top edge will, as stated, be exactly flush with that of the outer form and pitched at exactly the same angle. Use plenty of stakes with the smaller form, as it will

split the patio down the middle. Make its center high and its sides low. The same method suggested for constructing 20-foot-wide driveways in two sections in Chapter 7 can be used. Another solution is to break the patio into a number of separate sections with grass between the sections. Each section has its own pitch. When the patio adjoins your home, you can raise the adjoining patio end almost as high as you wish, as its height will be hidden by the building. When none of these suggestions work for your patio, you can always pile dirt up against the high side in a gentle slope. It won't be noticeable when grassed. In any case, it is inadvisable to reduce the pitch on a large slab. The reason is that it is difficult to keep the large surface perfectly flat; some shallow valleys are always formed. With little or no pitch to the slab, these depressions usually retain water.

Fig. 113. Poured-in-place concrete squares used to pave a large patio.

The second problem is that of pouring, screeding, tamping and floating any area much larger than 200 square feet within the allotted hour by an inexperienced three-person crew. The wiser course of action is to do the job in sections by dividing the slab into sections and, when necessary, by using stop boards where possible within each half.

PREPARING THE BASE

When and where you are going to lay down a granular stone base on which you will pour the concrete, use 1-inch or smaller stones. The larger stones are much more difficult to work with and do not pro-

vide any more support or drainage. Spread the stones as evenly as you can, bearing in mind that you don't want more than a 4-inch-thick slab. Thus, the surface of the spread stones must follow the pitch of the completed slab.

To check on the level of the base, place a board across the form and measure down. In this way you can be certain that no portion of the slab will be less than 4 inches thick, and that you will not be wasting concrete filling a depression.

To save on the crushed stone or gravel, you can fill the low areas with fieldstones, bricks, old concrete and concrete block. When you do, keep the surfaces of these fillers below the under surface of the concrete you will pour. And, be certain to hose them all down to remove adhering dirt and to thoroughly wet the porous bricks and concrete. If you don't, they can draw water from the fresh concrete and cause it to harden prematurely. Make certain the granular base goes beneath all the form boards and beyond for a couple of inches. You need this spread to provide a firm support for the edges of the slab. You also need to close the openings beneath the boards so that the concrete does not run out.

Where the earth slopes more steeply than you need, you will have to build the low area up with stone (Fig. 114). But no matter how much stone you may pile up, the stone cannot rest on grass. You must remove all the grass and its roots, even though you may have a foot or two of stone piled up at the low end of the patio form. Once the slab is in and the form is removed, you can cover the exposed stone with sod.

MAKING THE SLAB

Use the high-sand mix, the 1:2¼:3 mix suggested for walks and driveways. Mix or have it mixed a

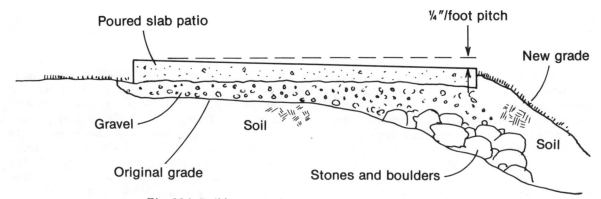

Fig. 114. Building up a slope to support a near-level patio.

Fig. 115. Screeding along the side of a large concrete patio.

Fig. 116. Using a commercial bull float to help finish a large slab.

Fig. 117. Floating the edge of a large slab. Courtesy Goldblatt Tool Co.

Fig. 118. Steel-troweling the same slab. Courtesy Goldblatt Tool Co.

little on the wet side, and use stones less than 1 inch in size, as there is no need for larger ones.

Single sections. Follow the same procedure for walks and drives. Fill one end of the form with concrete. Puddle it a little with a shovel or rake to help it settle into position and screed forward. Pour some more and screed some more. On a hot day, don't let more than a couple of yards of screeded concrete lie there without tamping. Repeat until the whole form is filled. Edge all sides and float-finish the surface as soon as it has set up. Cover the concrete so that it will not dry out, but will cure.

Should you want a smooth finish, follow your float with a steel trowel (Fig. 118). Use sweeping, circu-

lar motions, with the leading edge of the trowel raised a bit to keep it from digging in (some masons sprinkle a little water ahead of the trowel to help it along).

Note that a floated surface is smooth but grainy. It has some friction, and is the best surface for walks and drives. Following the float with a steel trowel makes the surface even smoother and reduces its frictive qualities. Steel troweling is not recommended for walks and drives. This finish is used on level surfaces which, like cellar floors, are normally dry, and patios not used when it's raining. Also, floated finishes do not take paint well. If you plan on painting your patio, you should steel-trowel it. Note

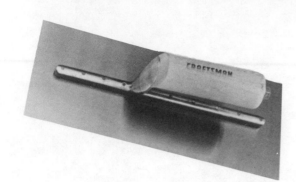

Fig. 119. A steel trowel.

that steel troweling takes as much or more time than floating, which means that your available work time before the concrete begins to set is further reduced when you steel-trowel. Thus, you should pour even less concrete when you plan on a trowel finish than when you plan to just float-finish the slab.

Multiple sections. Pour and finish the section farthest from the source of concrete so that you will not have to walk across the completed slab in order to continue working. If you are using a stop board along the length of a form, you can remove the stop board and continue as soon as the first batch of concrete has been finished. In such cases, your screed will rest on form boards alone. Where you have to rest your screed on a concrete slab, you had best wait several days, otherwise you may damage it by removing the form too quickly and dragging the lumber over the still green concrete.

COLORING CONCRETE

There are several ways of coloring concrete, detailed as follows:

Integral one step. Dry color is added to the concrete as it is mixed. You need 5 to 10 pounds of color for each bag of cement. This is an easy, no-fail method, but the cost of the color can significantly increase the overall cost of the patio. Browns, greens and blues are available, with the latter two colors costing more than the others. For the greatest color intensity, use the maximum recommended color quantity; 10 pounds per bag of cement, as stated. For the brightest colors, use white sand and white cement.

Integral two step. This method saves a lot of color, but adds a lot of extra work. Construct the form as suggested, but make it exactly ½ inch lower all around than the desired final height. Pour and

screed the usual concrete mixture. Let the concrete set. Then carefully lift the entire form exactly ½ inch (use wedges and stones to keep it from moving downward under pressure). Mix up a batch of colored cement using white sand, white cement and the desired color (a ratio of one part cement to three parts sand is generally used), and fill in the last ½ inch now left in the form. Screed as you normally would. You now have a perfectly flat layer of colored mortar atop your concrete. Finish and cure in the usual way.

Dusting. After you have completed floating a concrete slab, the desired color is dusted over the still wet slab. Following, the slab's surface is again worked over with a float (and then a steel trowel if so desired). This method requires a minimum of color, but the color must be spread evenly over the concrete, and this is not easy. Should the color puddle, rest awhile and let it dry before continuing.

Staining. This is the easiest method of all: you simply apply the stain to the dry and cured concrete. Do not use over patched concrete, because even slightly different concrete mixtures will take the color differently.

Painting. Give the slab a month to dry and cure. Then hose it down thoroughly, let it dry and apply concrete paint.

FLAGGING A PATIO

Like the flagged walk described in Chapter 6, a flagged concrete patio begins as a concrete slab and is then covered with a layer of flagging set in mortar.

Flag thickness and excavation depth. Since you want the finished surface of the flagged patio to be 2 inches above grade, excavation depth will vary with the thickness of the flagging you select. Below is a table with some average dimensions.

FLAGGING THICKNESS AND DEPTH OF EXCAVATION

Patio composition	Depth of excavation
4″ concrete	2″
4″ concrete, 4″ stone	6″
4″ concrete, 1″ mortar 1″ flag	4″
4″ concrete, 4″ stone 1″ mortar, 1″ flag	8″
4″ concrete, 4″ stone 1″ mortar, 2¼ flag (brick)	9¼″

Flag size and patio size. The size of the flags is important only in that by varying your patio's dimensions to suit the flags you select, you can eliminate cutting. Allow ½ to ¾ inch for joints, and plan the slab so that the edges of the flags are flush with the edges of the concrete.

If you are going to lay the flags in a pattern (even random), it is best to spread the stones over your lawn first. Position them as you wish, then number them with water-soluble paint so that you will not lose track when you are actually placing them on the mortar.

Construct the slab. Proceed as suggested for building a bare concrete slab, but allow for the addition of mortar and flags on top. Float the slab but do not edge it. Give the slab two or three days to cure a bit. Then carefully raise the entire form a distance equal to the thickness of your flags plus 1 inch for the mortar (you can cut the mortar to ½ inch, but you will find working with it more difficult).

The top edge of the form is now your guide. Lock it in position with more stakes and stones to keep it from dropping while you work with it. When the job is done the surface of the flags will be flush with the surface of the form. When a flag's surface is uneven, its high point is made flush with the top of the form.

Prepare the mortar. When laying brick, concrete block and similar porous flags, use a mixture of 1 part mortar cement to 3 parts sand. When laying nonporous flags such as slate and flagstone, use regular cement and sand in the same proportions. To start, mix the cement and sand very thoroughly without any water, then add just enough water to enable you to form the mortar into a ball that will hold its shape without dripping. This is called a dry mix.

Laying the flags. Begin at any corner. With a brush, sprinkle a little water on 3 or 4 square feet of the concrete, and cover the moistened area with a layer of mortar 1 inch thick. With a trowel or small stick, level the surface of the mortar as best you can. Sprinkle a little water on the surface of the mortar, covering an area just a little larger than that which will be covered by the flag. Sprinkle a little water on the bottom of the flag and lay it in position. Tap the flag down with the heel of your hand.

Next, lay a straight-edged board across the form just above the flag. The entire top surface of the flag should just clear the underside of the guide. Also, the edges of the flag should be close to but not touching the form. If the flag is too high, remove it, remove some mortar, sprinkle more water on the underside of the flag and on the surface of the mortar,

and replace the flag. If it is just a little high, place a short piece of 2 × 4 on top of the flag and tap it lightly down with a small sledgehammer.

Continue flagging the slab this way until the entire surface is covered. To space the joints equally, use a short stick of the proper thickness. To keep the flags in line when you are not working alongside the form, stretch a line across the form as a guide. Do nothing to finish the joints at this time.

Finishing. Give the flags three or four days before proceeding to fill in the joints. Use the same dry mix suggested for supporting the flags. Sift the mortar carefully into the joints. Do no more than a foot or two at a time. Then, with either a brick trowel or a caulking trowel, press the dry mortar into the joint. Finish by sliding the trowel down the joint. This will compress the mortar and make it shine.

With a brush, carefully remove whatever mortar remains on the flags. If you do all this with sufficient care, the flags will remain clean. If the flags become soiled, try hot water and a scrub brush first, then apply a brick cleaner as directed. You can purchase these compounds at the mason supply yard.

Large patios. Complete all the sections of the patio by loosening and lifting the form the necessary distance, then lock it firmly in place by one means or another. Since you cannot place a long straight-edged board across the form (it is too large), you should use a mason's line stretched across the form.

Fig. 120. Rectangles and squares of concrete block of differing dimensions.

Fig. 121. Equal-size rectangles of concrete block.

Fig. 122. Hexagonal paving units.

As an alternative, you can position temporary guides on the concrete and work from them. These are simply long boards, placed on the concrete and raised to the proper height with wedges. The proper height, of course, is established by the mason's line stretched across the entire form. Bear in mind that the entire surface of the temporary guide board must be at the correct height, so check it along its length at several places, not just one end. To use, lay a board from the form to the temporary guide. When you have flagged up to the temporary guide, move it and continue. Be certain to keep checking its height, as it is all too easy to go up- and downhill on a large patio.

CHAPTER NINE

Building Permits

If you plan to erect a permanent structure larger than a doghouse within a municipality under the jurisdiction of a building department, you must first secure a building permit. If you build without a permit and seek permission following, you may or may not be fined, depending on local laws and assuming your work satisfies the *LOCAL* building code. If certain required inspections cannot be made because the work is already completed, such as inspection of the earth prior to pouring concrete for the footings, you will be forced to tear your building down. If visible portions of your work do not meet the code, you will be forced to rebuild to the code. Pressure is easily applied. The building department can apply a daily fine until you comply with their demands.

Who needs a permit? Just when building departments and zoning laws were introduced into the cities and towns of our country, no one has taken the trouble to record. They came gradually. Even now, some small communities are laying down local building codes for the first time. Builders were erecting ugly, dangerous structures with no concern beyond profit. Building codes and zoning laws have put and are putting an end to this.

To determine whether or not your property falls under the jurisdiction of a building department, check with your tax office and with the nearest city hall or town council. Since some areas are in a state of flux regarding building codes, make certain that you are or are not subject to the building laws, and that you know which building department supervises your property. If you are indeed under the eye of a building department, secure a permit before you start work; if you are not, you may build anything you like and place it anywhere you wish on your property.

PROCEDURE

To secure a building permit, you must submit a legal plot plan (map) of your property. The plan must show the existing structures, if any, and the proposed new structure (Fig. 123). In addition, you must also submit a complete plan of what you intend to construct. Both the plot plan and the building plan must be blueprinted, meaning they must be in permanent form so that they can be placed in the records. And, of course, there is a fee. There is always a fee.

Securing a plot plan. If you have just purchased a new building, a plot plan should be furnished by the seller. If you have purchased an old building, there is a chance the seller may have a plot plan in his records. If not, in some towns the building depart-

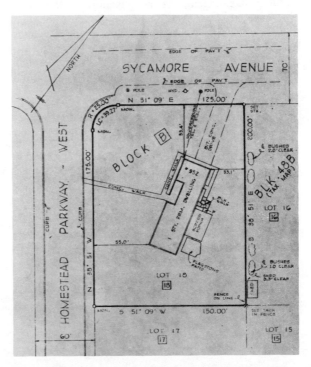

Fig. 123. Typical plot plan (surveyor's map).

ment will give you a copy of their plot plan for a fee.

If you cannot secure a plot plan from any of the above sources, or if you can secure a plan but there is no existing structure and the surveyor's stakes have disappeared, you must hire a licensed surveyor to survey, plot and locate your property lines. Note that you will not only need the plan to satisfy the building department, but that you will need a copy for a building loan application and one for your records. And you will need the surveyor's stakes from which to locate your structure.

With the plot plan in hand, a colored pencil may be used to indicate the relative dimensions and location of your proposed structure on the property.

Zone requirements. To be acceptable, your proposed structure must satisfy zoning requirements. These naturally differ from community to community and locality to locality. Essentially they are the same in that they work to organize buildings into industrial, business and residential areas. Your zoning board or building department will give you all the information you may require on this subject.

Code requirements. Code requirements are based on known good building practices. They vary slightly from community to community and building inspector to inspector. In some far-off century there will be one national code, but until then check with your building department. In no instance is the code difference between American communities so great that any building plan acceptable by one inspector cannot be altered to satisfy another inspector. In other words, where one town may specify 10-inch block, another will accept 8-inch block for the same wall.

Securing a plan. You can draw your own plan, purchase predrawn plans or hire an architect.

If you are going to draw your own plan, discuss it with the building department first. Generally, but not always, they are very helpful (they should be; you are paying their salaries). A well-made pencil drawing which is blueprinted for permanence is now generally acceptable.

Building plans of various types are offered for sale through the mail by many of the "home" and most of the "mechanics" magazines. Books of house plans in blueprint form are sometimes found on magazine racks. Most building departments will accept these plans without change. Others will demand specification changes which can easily be made on the existing plans. Still others will insist the plans be redrawn by a licensed architect.

Generally, if you look hard enough, you can find a licensed architect who will redraw finished plans for a reasonable fee. If not, you will have to hire an architect to draw your plans from scratch. This fee will be higher than if the man simply redrew the plans, but it now becomes his responsibility to make certain all local codes are met. If you wish, you can also engage the architect to supervise construction. In such instances his fee will range upward from 6% of the total cost of the structure.

Refusal and resubmission. Should your application for a building permit be refused, your next step (or your architect's next step) is to confer with the building department to learn what changes have to be made to make the proposal acceptable.

Variance. Should you wish to construct or position a building not acceptable to the building department, you can seek a variance. This is a legal procedure by which you can plead your case and possibly be granted the right to build counter to the code. For example, if your building lot was undersized according to the zoning laws you could try for a variance and so make it possible to build your home or whatever.

CHAPTER TEN

Layout

In present-day construction parlance, to *lay out* a building means to accurately position a proposed structure on its site by means of a number of stakes or pegs driven into the earth. Each peg marks a corner of the planned building.

This is much easier to do than it might sound. For one thing, a building may be an inch or two away from the exact position indicated on the plot plan and nevertheless be approved by the building department. Two inches is the usual acceptable error in layout. In other words, no one is going to scream if you build your garage or porch two inches closer to the property line than the code permits.

Additionally, the foundation may be an inch or two larger or smaller or even out of square without causing much problem. An inch or two one way or another is easily corrected between the bottom of the foundation and the building it will support. Thus, while accuracy is necessary, you don't need machine-shop accuracy. Anything within plus or minus an inch is acceptable.

LAYING OUT
AN ADDITION

At this point you should have three pieces of paper in your hand: a plot plan, showing the relative size and location of your planned addition, a plan of the addition itself and the building permit.

Use the existing building as a guide. If you are going to add a room to the side of the house, start your measurements from the building foundation itself. To lay out a rectangle or square adjoining the building, position the stakes with the aid of a steel tape. To make certain the stakes form right angles, measure diagonally from corner to corner. When the diagonals are equal in length, the square or rectangle has perfectly square—right angle—corners. To

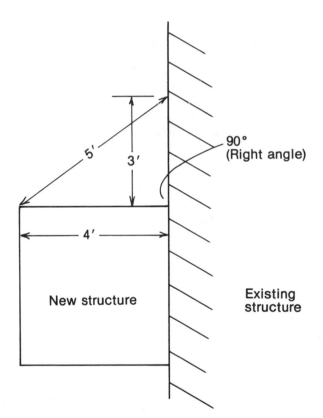

Fig. 124. *When the sides of a triangle measure 3, 4, 5 (inches, feet, meters, etc.), it is a right triangle.*

check on a single right angle, measure outward from the corner exactly 3 feet and mark the spot. Then measure outward from the same corner in the other direction exactly 4 feet. Mark the spot. Now measure from spot to spot, and the distance will be exactly 5 feet if the angle at the corner is a perfect right angle.

(69)

WORKING FROM SURVEYOR'S STAKES

Let us assume you plan to erect a building—any size or kind of building—on a lot having no accurately and legally positioned structures from which you can work. As in the previous example, you have in hand a plot plan and a plan of the proposed building. With the aid of the plot plan you can easily locate the stakes the surveyor has driven into the earth. The tack head on each stake is the surveyor's exact mark. The four stakes define the exact boundaries of your property.

Laying down the side lines. For simplicity and because most building lots are longer than they are wide, we shall assume that this is the shape of your building lot, and that one end terminates at a public road.

Your first step consists of demarking the sides of your property with mason's line (string). Begin at one side of the property and stretch a line over the two side stakes. Start and stop the line about 10 feet beyond each stake. Hold the line taut with two stakes of your own. Position it as close to the ground as you can without the line touching anything, running it exactly over the tack heads on the two stakes. To make certain this is so, hold a plumb bob alongside the line. The line will be directly above the tack head when the string supporting the

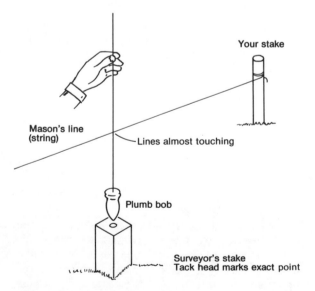

Fig. 125. How a plumb bob is used to position a line directly over a mark on a stake.

bob almost touches the line and the bob's point almost touches the tack head (Fig. 125).

If your lot is level, that is all you need to do to this side line at this time. If your lot is not level, you must raise the lower end of the side line until the line is level. You can tell when the line is level by hanging a line level at its center. When the bubble is in the center of the level, the line is level.

Now repeat the entire operation at the other side of your building lot.

Setback. Most codes require that a residence be set back a specified distance from the edge of the road. Generally there is no objection to a greater setback. Assuming that the front of your building will be set back this distance, measure along each side line the correct distance from the front stakes. Drive stakes into the ground and stretch another line across the width of the property at these points. This setback line should extend slightly beyond the two side lines.

Rear-of-the-house line. With the aid of a friend and a steel tape, measure from the setback line rearward a distance exactly equal to the depth (front-to-back) dimension of the house. Do this close to and parallel to one side line. Drive a stake into the earth to mark this measurement. Next, go close to the other side line, repeat the measurement and place a second stake there. Now stretch a line between the two stakes.

At this point you have two lines marking the sides of your property; one line parallel to the road marking the desired front edge of the planned building, and a second, parallel line marking the rear edge of the building.

Sides-of-the-house lines. Go to either side line and measure along the setback line in the direction of the other side line. Mark off a distance along the setback line that is equal to the space you wish to leave between that side of the proposed building and the side line of the property. Mark this distance on the setback line itself with a pencil. Do the same on the rear-of-the-house line. Next, with the aid of a helper, drive stakes into the earth and stretch a line over the two pencil marks you have made. This last piece of stretched string marks the desired position of one side of the building to be (Fig. 126).

Next, measure sideways along the setback line and the rear-of-the-house line and mark the spots that will indicate the opposite side of your building. In other words, measure and mark off the width of the building. This done, stretch another line over these two marks to locate the second side of your build-

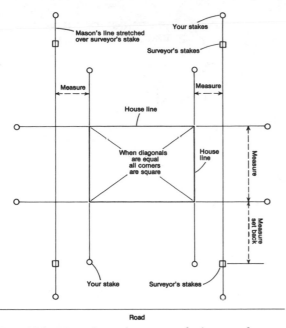

Fig. 126. Lines have been stretched over the surveyor's stakes. House lines have been measured off and stretched from stakes.

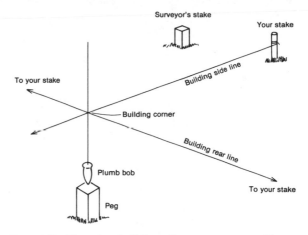

Fig. 127. Use the building lines as your guide to carefully peg the corners of the building.

ing. Now you can remove the side lines. However, do not move the stakes. You may need them again at some future time.

At this point you have four lines forming a square or rectangle that outline the exact size of your house and position it exactly where you plan to erect it. You cannot, of course, work with the lines in place. You have to remove them to excavate and build. To remove and replace them without loss of accuracy, and to do so quickly and easily, batter boards are installed and used.

BATTER BOARD INSTALLATION

You can use anything you wish for batter boards, even an old log lying in the correct place or a convenient tree. Let us assume the building lot is bare and you are going to make the batter boards from purchased lumber. You will need eight 5-foot-long roofers (the cheapest 1 × 6 boards) and sixteen 2 × 4s about 5 or 6 feet long. Point one end of each 2 × 4.

Peg the corners. The aforementioned lines cross at four points. These points are where you want the building corners to be. Drive a peg into the ground

beneath each crossover point (Fig. 127). With the aid of a plumb bob, make certain the top of each peg is exactly below its crossover point. Mark the exact point indicated by the tip of the plumb bob on the top of each stake. This accomplished, remove the four lines.

Position the batter boards. Each batter board is nailed to a pair of stakes, and its stakes are driven into the earth as shown in the accompanying illustration. Each board should be horizontal, on one plane, and about 10 feet clear of the nearest corner peg.

Replace the building lines. Stretch a line from the top of one batter board to the board facing it. With the aid of the trusty plumb bob, position the line exactly over the two corner pegs it crosses (Fig. 128). Install three more lines on the three remaining pairs of batter boards. Carefully mark the position of each line on each batter board. You will have a total of eight marks. Remove the four building lines. With a saw, make a shallow slot in each board right on the mark. Or drive a nail into the board just below each mark.

Using the batter boards. With the boards in place and marked, you have permanently located the four building lines (Fig. 129). Whenever necessary, you can quickly remove all the building lines and replace them just as quickly. Now you can excavate and work as necessary. Anytime you want to know where your building should be and just how large its foundation should be, you replace the four lines. That is all there is to layout.

(71)

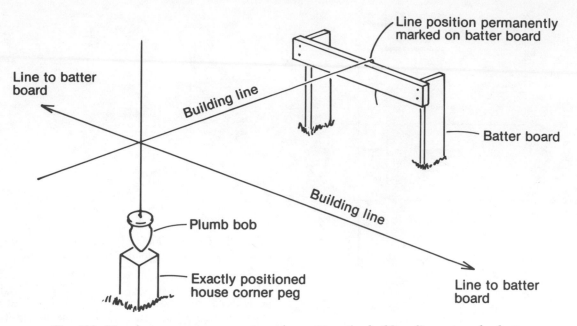

Line position permanently marked on batter board

Line to batter board

Building line

Batter board

Plumb bob

Building line

Exactly positioned house corner peg

Line to batter board

Fig. 128. *Use the corner pegs to accurately position the building lines atop the batter boards.*

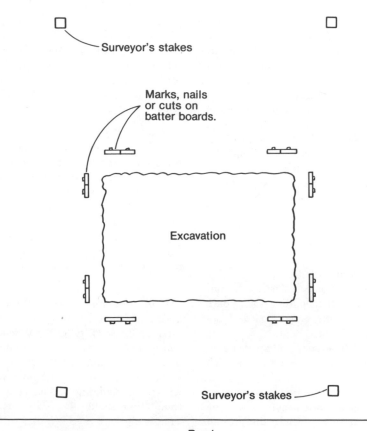

Surveyor's stakes

Marks, nails or cuts on batter boards.

Excavation

Surveyor's stakes

Road

Fig. 129. *Looking down on the building lot. Batter boards are positioned and marked. Anytime you want to know where any or all of the building lines are in relation to the excavation, you simply stretch the lines from board to board.*

CHAPTER ELEVEN

Footings and Foundations

Not too many generations back, builders would start by erecting four piles of stones on the ground and laying the floor girders atop the stones. In some parts of this country you can still see the corners of barns perched on crude stone piers.

We have since learned that a pile of stones atop the earth will not do. The stones shift with time and rise and fall with icy weather. For a stable foundation we must dig below the frost line and provide a solid masonry support. This is true of anything you may care to build, whether it is a porch, tool shed or a summer cottage or a mansion with thirty-six rooms. All structures must rest on a foundation, which in turn rests on firm soil out of the reach of frost.

TYPES OF FOUNDATION

Foundations can be grouped into three categories: pier, perimeter (Fig. 130) and slab. A pier is a column of masonry. Four or more are used to support a building. A perimeter foundation is a continuous masonry wall enclosing an area. When the perimeter foundation wall is low, the space between earth and the building above is called a crawl space. When there is more than sufficient room to stand erect in this space, it is called a cellar or basement. A slab foundation is literally a slab of concrete on which a building is erected. Slab foundations are discussed in the following chapter.

Pier vs perimeter. When a building is comparatively small and light, when there is no objection to an open space beneath it and when it is important to hold material and labor costs to a minimum, a pier foundation is used.

When it is necessary or desirable to have an enclosed cellar or crawl space, or when the building is to be made of masonry, a perimeter foundation is used.

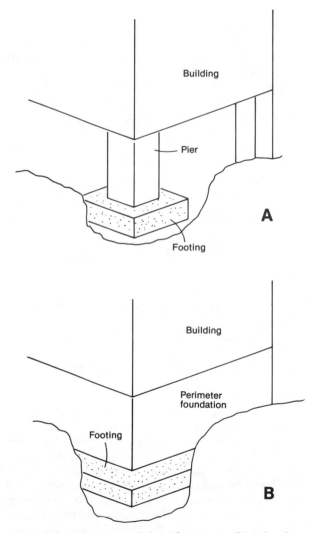

Fig. 130. Two types of foundation: A. Pier footing and foundation. B. Perimeter footing and foundation.

The difference in materials between the two can easily differ by a ratio of better than five to one. The difference in associated labor can run ten to one. That is why summer cottages, tool sheds and the like are almost always constructed on piers. Year-round homes and masonry garages are usually constructed on perimeter foundations.

FOOTINGS

A footing is a poured slab of concrete on which a foundation, pier or perimeter rests. The purpose of a footing is to prevent the foundation and the building from sinking into the earth. Since a building almost never subsides evenly, as little as 1 inch of subsidence can produce cracks in the building walls and make windows and doors difficult to operate.

Basic rules. Since frost heave can lift and damage a building as much as subsidence, footings are always positioned beneath the frost line. At a minimum, the bottom of the footing should be below the frost line, but for safety the entire footing slab should be below the frost line (Fig. 131).

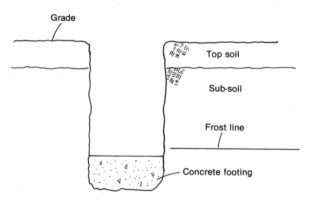

Fig. 131. Basic footing-depth rules. At a minimum, the footing must reach below the topsoil. If there is frost, the footing must be below the frost line.

Since topsoil is soft and contains disintegrating organic matter, footings should never be placed on topsoil. Instead, the earth should be excavated down to the light-colored, inert subsoil, even though there may be no frost at any time in the area.

Should you uncover bedrock at any depth, scrub the rock surface clean and erect the foundation directly on the stone. There is no need for a footing on bedrock.

If the building is comparatively light—a wood-frame cottage or smaller building is considered light

—and the earth is very hard and you are going to use a poured concrete foundation, you do not need a footing. If you are uncertain of just how hard the earth is, seek local engineering advice.

If your foundation is to be made of masonry, always provide it with a footing on all types of soils and beneath all types of buildings.

If the building itself is to be of masonry, always install a perimeter foundation and rest it on a footing on all types of soils.

Eliminating footing computation. To determine the minimum safe footing, it is necessary to test the load-bearing capability of the soil and then compute footing area according to the building to be supported. To safely eliminate all this, you can over-design. Make the footing larger than necessary even for soft clay, which is very soft soil. Should you encounter mud or silt, which is even softer, you will have to secure local engineering advice. Otherwise, the following rules of thumb will provide more than adequate support for cottages and similar structures:

1. Make perimeter foundation walls as thick or thicker than specified in your building plans, or specified by the local building code.
2. Make all masonry piers of pairs of blocks to provide a 16 × 16-inch cross-section.
3. Make all poured concrete piers 16 inches in cross-section.
4. Make all footings beneath masonry foundations twice as wide as the thickness of the wall and half as thick (Fig. 132). Thus, the footing beneath a wall of 10-inch block would be 20 inches wide and 10 inches thick.
5. Make the footings beneath all piers 3 feet square and 12 inches thick.
6. Make no pier higher than 5 feet from the top of the footing to the top of the pier without building department approval or local engineering advice.

CONSTRUCTING PIER FOUNDATIONS

First, inspect your building plans to make certain the building was designed to be supported by piers. Not all buildings are.

Second, lay out the building as previously shown. The four lines stretched between the batter boards should outline and position the frame of the structure before the siding is nailed into place.

Excavation. With a steel tape, rule and plumb bob, measure out along the building lines and locate the centers of the necessary piers. Next, dig the required holes to a depth of approximately 1 foot

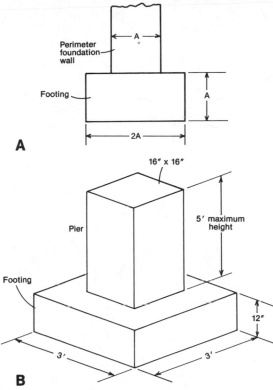

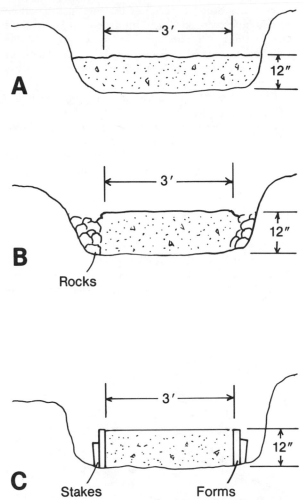

Fig. 132. A. Make a perimeter footing twice as wide as the thickness of the foundation wall it supports. Make the footing half as thick as it is wide. B. Make a pier footing 3 feet square and 12 inches thick. Make the pier 16 inches square and no more than 5 feet high.

Fig. 133. Three ways to make a pier footing: A. The hole is filled with concrete to a depth of 12 inches. B. Rocks are used to limit the spread of concrete in the bottom of the hole. C. A wood form of 2 × 12-inch lumber is constructed and positioned in the bottom of the hole. This form must be made level.

below the frost line. Remove the lines if they are in your way; you can replace them easily enough. If the soil is firm, you can let the earth serve as a form for your footing. If it is not, you will have to construct a form within the hole if you do not wish to fill the entire hole bottom with concrete. There is actually no harm in doing this, except to your pocketbook. In any case, make the bottom of the hole as flat and as level as you can, roughly centered around the center of the pier. The footing's position is far from critical, as the pier can rest off center by several inches without problem. But, if possible, make all the footing holes the same depth below the building lines. In this way you can use an equal number of blocks to secure identical pier heights and will not have to cut any block. If you dig too deeply, replace the soil with rocks and stones. Do not throw loose soil back into the footing hole.

Footing forms. Use 2-inch boards on edge, nailed together as needed and held in place by stakes. If you wish, you can make rough forms out of piles of stones. You will need more concrete but you will save on lumber and labor. Just remember, you want to end up with at least a 12-inch-thick slab, 3 by 3 feet in size. A larger slab does no harm.

Pouring the footings. You may use any of the three mixes suggested in Chapter 2, but the foundation mix is the least expensive and will work fine. Before you pour, however, make certain the forms are level. Use a slightly wet mix. Screed the form after it has been poured. Let the mix in the bottom of the hole or rock "form" find its own level. Since the footings will not carry much weight until the building is erected, you can start to lay block or even pour a pier atop the footings shortly after it has reached initial set.

Block piers. Drop the plumb bob from the building line to locate the outside edge of the pier. A corner pier will have two "outside" edges, as shown (Fig. 134). Make the first course of two blocks positioned side by side. Use two blocks for the second course, placing them across the first pair, and so on up. Position the foundation bolt screw threads upward in the topmost block. There are the large bolts that hold the building in place. If you are uncertain after reviewing the plans as to the exact location of these bolts, let them be until you start constructing the building, at which time their exact position will become obvious. Fasten the bolts in place by surrounding them with mortar (discussed more fully later).

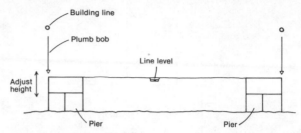

Fig. 135. *Use of line level to check relative elevation of two or more piers.*

Fig. 136. *Typical concrete-block pier.*

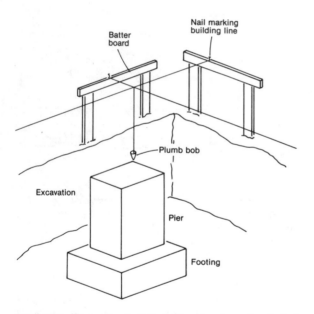

Fig. 134. *How the building lines and a plumb bob are used to locate a pier on its footing.*

The first pier you complete will establish the height of all the rest of the piers. Naturally, all pier tops must be on the same plane. To do this, stretch a line from the top of the completed pier to the pier under construction and hang a line level in the center of the line (Fig. 135). If necessary, you can raise the top of any pier by adding a layer of mortar or even 1-inch or 2-inch solid block on top.

Poured concrete piers. Make the forms from 1 × 6 roofers (tongue-and-groove yellow pine) and 2 × 4s. Lock the corners together by means of wire, metal straps or cross bracing. If you wish to take the forms apart quickly, lock their sides with hinges and hasps.

Form cross-section is not important so long as it is at least the recommended 16 by 16 inches. But form height is, as you want all the pier tops to be at the same level. Cover the inside of each form with motor oil so that they can be easily removed.

Position each form so that the concrete will be in line with the building lines. In other words, the cast concrete pier will be exactly where you would position concrete-block piers. Brace each form thor-

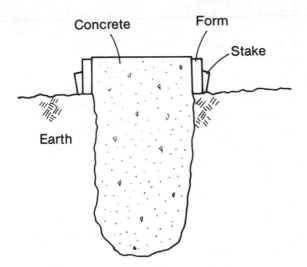

Fig. 137. Simple form that can be used to cast a concrete pier when the earth is firm and the pier is to be low.

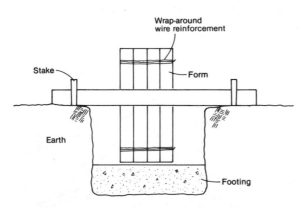

Fig. 138. Form that may be used when the earth is soft and the pier is to be high. Pier is poured after footing has set up.

oughly so that it will not move while it is being filled with concrete. Make certain you have the necessary foundation bolts on hand and you know exactly where they will go. Now check form elevations once again and you can pour.

Again, use any of the three mixes suggested in Chapter 2, but the foundation mix will do fine. Use a little less water than usual. Pour very slowly. Poke the mix from time to time to make certain it settles properly. When the form is full, insert the foundation bolt, threads up. You can pour all the piers at one time if you wish, or you can permit any period of time to elapse before you pour the concrete into

the form on top of the footing. Just make certain the top of the footing is perfectly clean before you pour. Dust and dirt will prevent proper bonding between the fresh concrete and the old.

CONSTRUCTING A PERIMETER FOOTING

Essentially, all perimeter footings are alike. They differ only in dimensions and in the depths at which they are positioned.

General excavating practices. Since a footing is never laid down on topsoil, and since the area within a perimeter footing is almost always covered by a slab of concrete or a layer of blacktop (an asphalt and stone mixture), the first step is always to remove the topsoil. Generally, the topsoil is removed from an area larger than that to be occupied by the building. The removed topsoil is formed into one or more piles of earth well clear of the work area.

The earth that is removed to make room for a full cellar, crawl space or just a footing trench is collected in a separate pile or piles a good distance from the work area. In some instances you may want to or need to haul this dirt away, but most often you will need this dirt to back-fill (fill the spaces alongside the foundation) and to provide an uphill pitch toward the building. The uphill pitch is very desirable, as it forces surface water to flow away from the building. By utilizing the excavated soil around the building, you will also save yourself the cost of hauling the soil away.

When excavating for a full cellar, slope one side of the excavation to make a driveway down which trucks, wheelbarrows and materials can be brought into the center of the job. The closer concrete block and other materials can be brought to where they will be positioned, the less physical work required. As the average small building cellar requires some 30 to 40 tons of block alone, just moving the block by machine to where it will be needed can save a tremendous amount of time and labor.

Formless footings. Stretch the building lines. Mark the building corners with small pegs driven into the earth. Measure outward from each peg to define an area 5 feet larger all around than the space the building itself will occupy. Remove sufficient topsoil from this area to uncover the subsoil. Move the topsoil out of the way.

Now, assuming that the soil is sufficiently firm and that you do not have to dig too deeply, replace the building lines and corner pegs. Using the building lines as your guide and a plumb bob to transfer the position of the lines onto the earth, scratch the outline of the needed footing trench on the surface of the subsoil. Bear in mind that the outside edge of the footing extends several inches beyond the edge of the block, which will be positioned directly below the building line.

Dig the trench. Make its bottom as flat and level as you can. Measure down from the horizontal building lines to check on trench-bottom depth and levelness. Make the bottom of the trench just as wide as you want your footing to be. Should soft soil force you to slope the sides of the trench, no problem. You will just require a little more concrete.

Install service pipes and lines. All the pipes and power lines that must pass beneath the footing should be installed now. Once the footing is in, you will have to tunnel underneath, and that is very difficult. Simply dig a narrow trench across the footing trench. Install the drainpipe and power lines. Cover the pipes and lines with soil and go ahead with the footing. A narrow strip of "soft" soil beneath the footing will not do it any harm. Pipes and lines that go through the foundation wall are installed when the wall is erected.

Pouring formless footings. If you are going to pour concrete footing directly into a trench, you need no form (Fig. 139). But the footing trench you dig may need some preparation. First, check to make certain the trench is positioned correctly in relation to the building lines. Then make certain the bottom of the trench is as deep as it needs to be and as wide at the bottom as it needs to be. The extra footing width that will result because of the sloped walls of the trench means a little extra concrete and adds a little strength, nothing more.

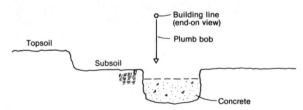

Fig. 139. When the earth is firm, the footing can be poured directly into a trench dug in the earth. Note position of building line.

Remove all the loose stones and clods of earth that may have fallen inside. Prepare a 1:3:5 mix on the wet side, wet enough to flow like thick soup. Fill the trench very gently from as many different points as is practical. This will reduce to a minimum the

amount of pushing and pulling needed to position the concrete. If you agitate the concrete too much, dirt may fall into the trench and mix with the concrete. This will weaken the footing. Let the concrete find its own level. Even if it becomes too thick to lie perfectly flat, a difference of even as much as 2 inches end to end can be corrected when you lay up the block. Check footing thickness by either measuring down from the building lines or simply poking a ruler through the concrete. A variation of an inch or so in the thickness of the footing is unimportant.

Formed footings. The difference between a formless footing and a formed footing is that, as described, the formless footing is made by pouring concrete into a trench while the other is made by pouring concrete into a form. Most often, the footing form is constructed of 2-inch boards, set up on edge and held in place by stakes (Fig. 140). Every dozen feet or so a brace is nailed across the form to hold the sides together. Since the lumber is only soiled by its use as a footing form, these boards are often incorporated into the building as joists and parts of the girder.

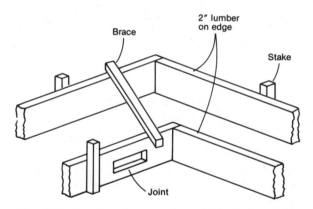

Fig. 140. Footing form made of 2-inch boards set on edge, joined end to end and staked and braced as shown.

The same procedure used to locate and dig the trench necessary for a formless footing is used when a form is to be constructed. Obviously, the trench must be that much wider, but the same relationship between the building lines and the outside edge of the finished footing exists as before.

Make certain one side of the form is level with the other by placing a spirit level across it. To make certain one end of the form is level with the other end, stretch a line from end to end and hang a line level

Fig. 141. Typical small building lot prior to excavation. Pairs of batter boards have been joined, and a white line of lime laid down to guide the bulldozer.

Fig. 143. The point at which building lines cross will be where the corner of the building will be erected.

Fig. 142. End view of footing form. Note position of building line in relation to form.

Fig. 144. Ready-Mix concrete coming down chute is directed into footing form. Iron rake is used to pull concrete along.

in the center of the line. Do not depend on a spirit level placed on top of a board edge. It will lead you astray.

Pouring a formed footing. Double-check to make certain the form is properly positioned beneath the building lines. Check to make certain the form sits firmly on the earth, that the sides of the form are level, that the entire form is level end to end and

side to side. See that there is a stout stake holding the form down every 10 feet or so, and that a brace is nailed across the form every 12 feet or so. Fill all spaces between the bottom of the form and the earth with stones. Then pile earth atop the stones from the outside of the form. The earth will help keep the concrete from leaking out beneath the form.

Use the 1:3:5 mix suggested previously. Add

(79)

Fig. 145. Form has been filled. Rake is used to "vibrate" concrete to make certain it settles into all corners of the form.

Fig. 146. Another view of the same job. Note that block has been brought into center of excavation where it will be easily reached by the masons.

sufficient water to enable you and your helpers to push the concrete into place. Then, with a short, straight stick, screed the top of the concrete flush with the tops of the form boards. If you cannot pour the entire form in one day, pour and screed a section and poke 3-foot-long, ½-inch-thick steel reinforcing bars into the end of the in-place concrete. Four bars should be enough. When you resume pouring, the steel bars will help lock the new concrete to the old.

Give the pour two days to set up and harden. Then carefully remove the form. If you wait too long, you will have a devil of a time separating the boards from the concrete.

Excavating a full cellar. Before you begin excavating, decide how high you want your cellar foundation wall to extend above the finished grade. Decide how many courses of block will be necessary to provide the desired finished cellar ceiling height when the pipes are in and the concrete cellar floor has been poured (this goes on top of the footings). Decide the necessary thickness of the footing, and decide whether or not you are going to construct a formed or formless footing.

If you are planning to construct a footing form, you can position it directly on top of the dirt cellar

floor, in which case you will have to cover the entire cellar floor with a layer of crushed stone or gravel as high as the top of the footing form. Or you can place the form in a trench. If the top of the footing form is made level with the surface of the dirt cellar floor, you can pour the floor directly on top of the soil. This is commonly done when no cellar water is expected. If you want to lay down a 4-inch layer of gravel, position the top of the footing form 4 inches higher than the dirt cellar floor. All this is perhaps more clearly explained in the accompanying illustrations (Figs. 147 and 148).

If you are not going to use a form, simply dig the necessary trench in the dirt cellar floor. The top of the poured footing is then flush with the top of the dirt floor. The concrete cellar slab then goes on top of the dirt and the footing.

Start your excavating by removing the topsoil from an area 20 feet longer and 20 feet wider than the building that is to be erected. Then excavate a hole sufficiently larger than the foundation you are going to build to give you at least 2 feet of clearance alongside the footing. You need this space to do your work and possibly to install drainpipes.

Make the bottom of the excavation as flat and as level as practical. To do so, you will need to check

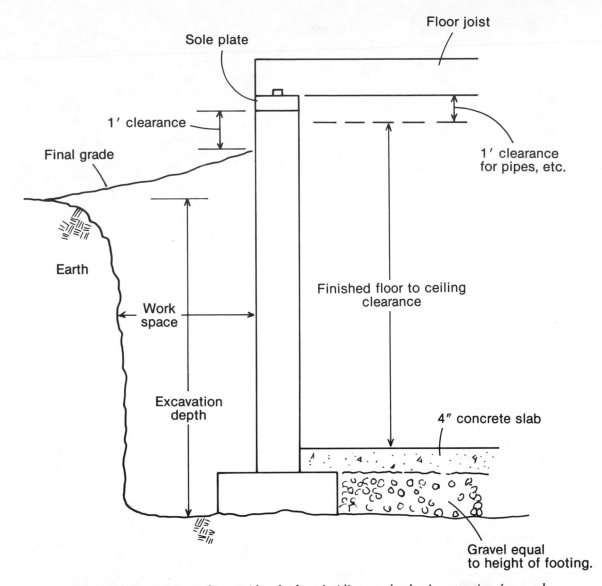

Sole plate

Floor joist

1′ clearance

Final grade

1′ clearance
for pipes, etc.

Earth

Work
space

Finished floor to ceiling
clearance

Excavation
depth

4″ concrete slab

Gravel equal
to height of footing.

Fig. 147. Dimensions to be considered when deciding on depth of excavation for a cellar or crawl space. Note that the footing rests on the bottom of the excavation, which makes it necessary to fill the space between the soil and the top of the footing with gravel.

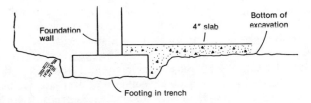

Foundation
wall

4″ slab

Bottom of
excavation

Footing in trench

Fig. 148. Here the excavation is made shallower by the thickness of the footing, which is placed in trench. No gravel is needed.

(81)

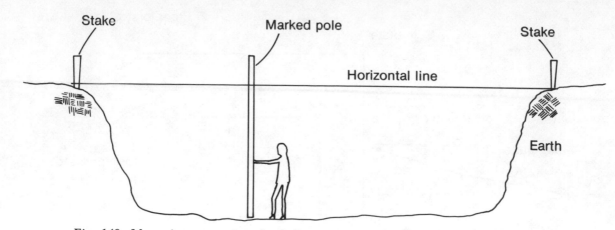

Fig. 149. *Measuring excavation depth by means of a marked pole and a horizontal line.*

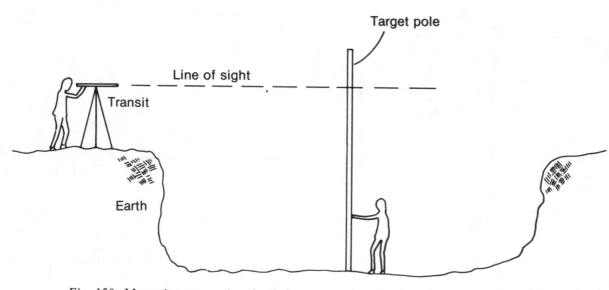

Fig. 150. *Measuring excavation depth by means of a transit and target pole (marked stick). With scope on a horizontal plane, everything that can be seen through the scope is on the same plane.*

excavation depth from time to time as you go down. One way consists of replacing the building lines, which are horizontal, and measuring down from them. To measure depth of areas not under the building lines, stretch another line across a pair of building lines. Use stakes to hold this line in position, and elevate it so that it just clears the building lines. To simplify measurements, use a long stick with the desired depth marked on its side. In this way you can avoid stopping and reading numbers on a rule each time you check excavation depth.

Using a transit. Measuring down from a horizon-

tal line stretched across the excavation is sufficiently accurate for our needs, but it is far too slow a method when a bulldozer is waiting. The time wasted while you or your helper measures excavation depth with a stick will cost you more in dozer rental fees than the rental of a transit or even the purchase of a low-priced unit. Even the cheapest transit is more sufficiently accurate for our work (the best, called a theodolite, is accurate to plus or minus 0.10 inch in a mile).

When you believe you are down to a foot or so higher than the desired depth, station your transit at

(82)

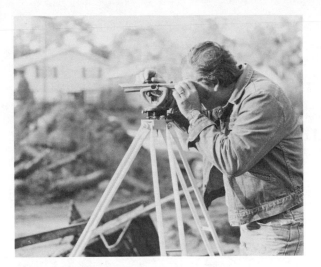

Fig. 151. Using a transit.

any point alongside the excavation from which you have an unobstructed view of the entire job (Fig. 150). Adjust the transit platform until all the bubbles in the platform spirit gauges indicate the platform is level. Adjust the sighting telescope until its gauge indicates that it is level. Then rotate the sight through 360 degrees and watch its bubble. If the supporting platform is perfectly level, the sight will remain perfectly level. If it is not, correct it before going any further. Now measure from the center of the horizontal sighting scope to the earth. Let us assume this distance is exactly 3 feet. Have a helper hold a target pole erect, its lower end resting on the earth within the excavation. If you do not want to rent or purchase a target pole, use a long, straight stick in its place. Mark it this way: if the scope is 3 feet above grade, add the desired depth of excavation to this figure and mark the total distance from one end of the stick. Assuming you want to dig down 6 feet and the scope is 3 feet aboveground, your pole mark would be 9 feet from one end. Now sight through the scope at the target pole. Adjust the focus knob until you see it clearly. If the mark is above the cross hairs, you still have to dig down. If the mark is below the cross hairs, you have dug too deeply.

The transit can also be used for leveling footings and walls.

To use the transit for leveling footing forms, position it on the dirt cellar floor and level it. Then have a helper hold a ruler erect at any point on the form. Position the ruler so that the numbers read from zero at its lower end on upward. Sight the ruler.

Note the number in the cross hairs. Have the helper move the ruler to another position on the footing form. Take another sighting. If you see the same number, the two points are on the same plane and are level with one another. If you see a higher number, the second point is lower than the first by the difference in numbers, or vice versa. The same method is used when you want to check the elevation of a foundation wall under construction, or the relative elevation of any two points within the sight of the scope.

Lally column footings. A lally column is a steel tube filled with concrete. It rests on a small steel plate and is topped by another small steel plate. It is used to provide vertical support for girders. Its position and size should be marked on your building plan. It rests on a footing that is usually 3 by 3 feet and just as thick as the perimeter footing. The top of the lally column footing is almost always flush with or slightly below the surface of the perimeter footing. Generally, the lally footing is poured when the perimeter footing is poured.

If you are not going to build a form, simply dig the required hole as deep as the perimeter footing trench and fill the hole to its top with the same mix as used for the perimeter footing. If you are using a form, use material of the same width as the perimeter footing form. Don't worry too much about its size or placement; the lally column does not have to be too accurately positioned. But do take great care to keep the top of the lally footing flush or lower than the top of the perimeter footing. Note that the bottom plate and the column are both positioned just before the concrete cellar floor is poured. This floor locks the column and its support in permanent position.

Crawl space footings. The major difference between cellar and crawl space footings is the greater depth of the full cellar footing, nothing more. The minor difference is usually in the manner in which the exposed earth is covered. In a cellar, a 4-inch, load-bearing, perfectly smooth and level floor is laid down from foundation wall to foundation wall. In a crawl space, no more than 2 inches of concrete or blacktop is ever used. The surfaces may or may not be smooth or level; their purpose is to provide a moisture barrier, to keep the building from absorbing moisture from the earth. Most often, this barrier terminates at the sides of the footing. Of the two, the blacktop is the easier to use. All you need do is spread it with a rake. It will of itself follow irregularities in the earth and seal itself to the foundation footing.

CHAPTER TWELVE
Slab Foundations

As stated, a slab foundation is essentially a slab of concrete which rests on the ground and supports the walls of a building. Its advantage is that it serves two functions: it is both a foundation and a floor. When and where the slab is suitable, it is the least expensive and the quickest way to construct a building. It may be used to support both masonry and wood-frame structures.

BASIC DESIGN

Simple slab. When and where there is no frost, the earth is firm, hard and dry, and the building is only one story high and relatively small, the slab can be laid directly on subsoil (Fig. 152). A one-story building of masonry no more than 20 feet wide can be carried on a 4-inch slab. Wood-frame buildings can be built to a width of up to about 26 feet. Larger buildings require thicker slabs. The simple, laid-on-the-ground foundation is excellent for sheds, cabanas, garages, pool buildings—any structure that does not have to be more than a few inches above grade.

Thickened-edge slab. When the weight of the planned building is too great for a 4-inch slab, the slab has to be thickened (Fig. 153). When the building is to be of wood and the climate is temperate, the top of the slab should be at least 6 inches above grade. When the climate is semitropical or particularly wet, the top of the slab should be 12 inches above grade to protect the wood against rot and insects.

A 12- or even a 6-inch-thick slab requires considerably more concrete than a 4-inch slab. But, as the need for the extra thickness is only along the edges of the slab, the solution lies in constructing a 4-inch-thick slab above a suitable layer of crushed stone and just making the edges of the slab as thick as

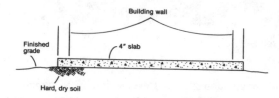

Fig. 152. Where the ground is hard, dry and frost-free, a small building can be constructed on a 4-inch slab placed directly on the earth.

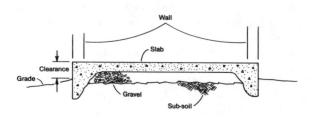

Fig. 153. Cross-sectional view of a thickened-edge slab foundation. Thick edge carries weight of building and also provides greater above-grade clearance.

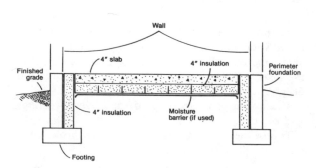

Fig. 154. Here the slab is insulated from both the earth below and the perimeter foundation wall.

necessary. Thus, the cost of concrete is greatly reduced.

Insulated slab. To prevent heat from leaking out of the building through its concrete slab floor and integral foundation, the slab and its thickened edge may be insulation from each other and the earth, separated with rigid polyurethane or glass bubble insulation (Foamglass) (Fig. 154). In such cases, the slab is usually constructed as a poured unit within a poured or laid-up block perimeter foundation.

CONSTRUCTING A SIMPLE SLAB FOUNDATION

As outlined previously, if you are within a controlled municipality, secure a building department permit before doing anything. Not only will you avoid difficulty with the authorities, you will assure yourself that your design is structurally sound.

Laying out. Proceed as directed in Chapter 10. Install the batter boards. Stretch the building lines. Peg the corners of the building to be. If there is a layer of topsoil, remove it to make an area 5 feet longer and wider than the slab is going to be. If the uncovered soil is still soft and wet, you should excavate 4 more inches or so and lay down a layer of crushed stone or gravel as suggested in Chapters 6 and 7 for walks and driveways on soft, wet soil.

Constructing the form. Build the form of 2 × 4s on edge. Remember that the inside of the form (the finished slab's dimensions) must be exactly the same as the area demarked by the building lines. When you drop the plumb bob from the building line, the bob's tip should just touch the inside edge of the form.

If you are going to bring any service pipes and lines into the building through its floor (slab) now is the time to install them. When positioned, wrap several layers of tar paper around the pipe or cable where it will pass through the concrete. The paper insulates the pipe from the concrete and lets the pipe move a little in response to temperature changes and the like.

If the slab is going to support a garage, adjust the form so that it pitches ⅛ inch per foot to where the door will be. If not, make the form as level as you can. In the case of a small form, you can lay a long, straight-edged board across the form and place a spirit level on top. If you do not have a straight-edged board long enough to span the form, use a line and a line level.

When and where the ground is pitched, you have the choice of either excavating a portion of the earth or raising the form. When you raise the form, fill the space with crushed stone or gravel and spread the gravel five or six inches beyond the form for stability.

Pouring and finishing. If you want a smooth finish on the slab, use a 1:2¼:3 or a 1:2¾:4 mix. The high-sand mixes will respond better to steel-troweling. If you are merely going to float-finish, the 1:3:5 mix will do fine. Pour, screed, tamp and finish as suggested in Chapter 8.

Installing the foundation bolts. If you are going to place a wood-frame building atop the slab, install foundation bolts head down, in the still-soft concrete. Position them so they are centered in the building's sole plot and not beneath a door or a stud—one bolt every 6 to 10 feet or so will do.

CONSTRUCTING THICKENED-EDGE SLAB FOUNDATIONS

When there is a chance of frost and/or when a two- or three-story building of concrete block is to be erected on the slab foundation, the edge of the slab must be thickened to go beneath the frost line and to carry the weight.

There are two general ways to build this type of slab foundation. One method consists of raising the entire slab (Fig. 155). To do so, the circumscribing form is made high, say a foot or so. Then, to avoid pouring a foot-thick slab, the central area of the form is partially filled with crushed stone. When the form is poured and screeded, the edges of the slab will be 1 foot thick, while the balance will be much thinner because of the stone.

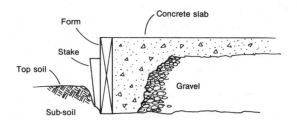

Fig. 155. How a high form may be used to produce a thickened-edge slab. Gravel pile reduces concrete thickness of the middle of the slab.

The other method utilizes a normal height form (Fig. 156). To secure the thick slab edge, a trench is dug alongside the inside of the form. When this form is poured and screeded, the resultant slab has

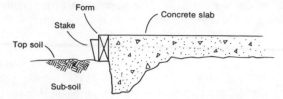

Fig. 156. Here a trench has been dug along the inside edge of the form. The trench provides the extra slab thickness at its edges. No gravel is needed.

thick edges because the concrete goes down into the trench. This method is used where frost is expected.

High-wall form construction. Remove the topsoil. Construct the form from boards just as wide as you wish the top of the slab to be above grade. For example, you have removed 6 inches of topsoil (which you will eventually replace). You want your slab top to be 6 inches above "finished" grade. To do so, you would use a 2 × 12 set on edge. This would result in a slab having an edge 12 inches high.

With the form completed, stake it in place securely. Stretch the building lines from the batter boards. Drop the plumb bob from the crossover points of the lines, which locate the slab corners. Check to make certain the inside form corners are directly below the corners indicated by the crossed building lines. Make certain the form is level.

This done, fill the central area of the form with gravel or crushed stone. Make the top of the stone reasonably level, 4 inches below the top edges of the form. Keep the stone 1 foot away from the inside edges of the form. Thus, you will end up with a low, flat-topped mound of stone within the form but clear of it by a foot or so all around.

Now you are ready to pour. Carefully fill the open space between the stones and the form first. Do so slowly so as not to apply too much pressure to the form. When this area is filled with concrete to the level of the stone, you can fill the balance of the form. Screed, tamp, float and steel-trowel, as usual.

If your foundation slab is too big to screed with one pass, divide the slab into sections. Pour and screed one section at a time. Then remove the dividing guides. Fill the spaces with concrete. Screed the fresh concrete flush with the adjoining concrete and finish as usual.

Low-wall form construction. Remove the topsoil, build a form of boards of a width equal to the desired slab thickness. Position the form beneath the building lines, as previously described. Make the top of the form as level as you can. Next, alongside the

inside edge of the form, dig a trench at least 6 inches wide and at least 1 foot deep. Where there is frost, the trench bottom should be below the frost line. When 10 or so vertical feet of block are to be laid atop the completed slab, the trench should be at least as wide as the thickness of the block. When two stories of block are to be laid atop the slab, the trench should be at least twice the thickness of the block. For example, beneath 8-inch block, two stories high, trench width should be 16 inches at a minimum.

Carefully pour the concrete into the trench before you fill the balance of the form. Then screed, tamp and finish as before.

INSULATED-SLAB CONSTRUCTION

The insulated slab is constructed in two sections. The first is actually a perimeter foundation wall. The top of this wall determines the top surface height or elevation of the slab. The wall is insulated from the soil on the inside by rigid batts of insulation. These batts also insulate the wall from the slab. The slab is poured on top of a layer of insulation, which rests atop a moisture barrier, which in turn rests on crushed stone. The result is a slab that is insulated from moisture and cold that may come up from the earth below, and insulated from cold that may come in from the foundation wall.

Excavation. Start by removing the topsoil for a distance of 5 or 6 feet beyond the building lines. Dig a perimeter trench 2 feet wide. Make the bottom of the trench as flat and level as you can and at least a few inches deeper than the frost line. Position the trench so that its exterior walls are approximately 1 foot beyond the building lines.

Footing. Either pour the bottom of the trench full of concrete or construct a footing form within the trench and pour it full of concrete. Follow the suggestions given in Chapter 11 for constructing footings.

Perimeter foundation. Construct a low-wall, concrete-block perimeter foundation on the footing (wall construction details are covered in Chapter 13). Stretch the building lines and drop the plumb bob to locate the corners on the footing. Make the foundation exactly as high as you wish the top of the slab to be; the surface of the slab, when poured, should be flush with the top of the foundation wall.

Insulate the foundation wall. Place batts of 4-inch thick, rigid Styrofoam or similar insulation up against the inside surface of the foundation. Position the tops of the batts flush with the top of the wall.

Butt the side of one batt tightly against another. Cut the batts with a razor knife as necessary so that nowhere does one batt overlap another. Toss stones and rocks into the spaces between the batts and the soil to lock the batts in position.

Preparing for the slab. Cover the earth within the area enclosed by the foundation wall with 6 inches of crushed stone or gravel. Spread the stone out evenly. Check on its height by measuring down from a board or string laid across the area from foundation wall top to foundation wall top.

With a metal tamper, pound lightly on the stones. Your purpose is to blunt or turn all the sharp stone points so they will not pierce the moisture barrier that will rest on top.

Cover the stones with a sheet of 4-mil or thicker polyethylene. Cut the sheet so that its ends meet the wall squarely and there are no open spaces for moisture to come through. Overlap the sheets by a foot or more when one sheet will not cover an entire area.

This done, cover the entire area with 4-inch-thick batts of rigid insulation. Once again, make certain the batts butt tightly against one another and the encircling walls. Cut the batts as necessary so that none overlap or leave openings.

Pouring the slab. At this point you have a 4-inch-deep, square-sided "pan" formed by the vertical insulation behind the foundation and the insulation lying flat on top of the moisture barrier and crushed stone (Fig. 157). The concrete is poured into this pan and screeded level with the top of the foundation wall, then tamped and finished as usual.

Poured foundation. Instead of constructing a foundation wall of block resting on a poured footing, you can, if you wish, make the wall in a single pour. This is accomplished with the aid of a wall-high form of plywood positioned within the aforementioned perimeter trench (Fig. 158).

Make the form from ¾-inch construction-grade plywood nailed to 2 × 4 braces. The braces are positioned vertically and backed by horizontal braces called *walers*. When the form is 2 feet or less in height, you will need one brace every 3 feet. When the form is 3 feet in height, use one brace every 2 feet. Use two walers, fastened on edge, on both sides of the form. Space the sides of the form 1 foot apart, which, of course, will produce a wall of this thickness.

To keep the sides of the form from moving apart under the pressure of the concrete, drill holes through the plywood. Loop heavy galvanized wire around the walers and through the form. These

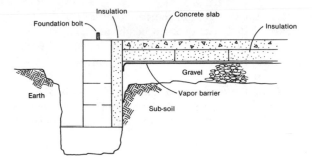

Fig. 157. *Insulating batts are placed on top of the moisture barrier. Concrete is poured atop the insulation. The concrete is screeded flush with the top of the foundation wall.*

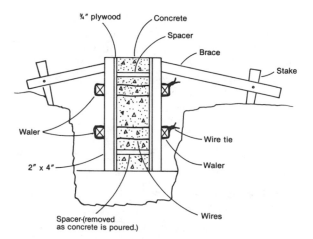

Fig. 158. *End view of wood form that may be used to cast a poured foundation wall. Note that footing has already been poured and has set up.*

wires remain in the concrete; later, cut their edges flush with the wall. Use braces to hold the form in position and to keep it from rising when the concrete is introduced.

Check the position of the form against the building lines. You want the outside of the completed, poured concrete foundation to be directly below the building lines. Make certain the top of the form is perfectly level, for you will screed the pour flush with the top of the form. Make certain the top of the form is just as high as you want it to be. Remember, the crushed stone, insulation and slab total 14 inches. Should you change any of these dimensions, you would have to change the height of the form top, and vice versa.

Use a 1:3:5 mix a bit on the wet side. Pour or

place the concrete in the form very, very slowly, no more than 1 foot of form height per hour. If you fill the form too rapidly, you will burst it. With a straight-edged stick, screed the concrete flush with the top of the form.

Give the concrete a day or two to harden and then remove the form very carefully. The procedure from here on is exactly the same as previously described.

SLABS FOR FRAME BUILDING CONSTRUCTION

Since wood rots and is subject to termites and other wood-eating creatures, it is advisable to make the top of a slab that is to carry a wooden structure of any kind at least 6 inches above finished grade (when the topsoil is in place). In warm, wet areas make this clearance at least 12 inches.

When you build a concrete or brick wall atop a concrete slab, all that is necessary to bond the masonry to the concrete is a layer of common mortar.

Obviously, this will not work with wood. When you expect to construct a wood-frame building atop a concrete slab, you must install foundation bolts in the concrete.

The bolts are positioned head down, threads up, about one every 6 to 10 feet, as described earlier. Holes are drilled in the sole plate, which is laid over the bolts. Nuts lock the sole plate down to the slab.

Use 10-inch bolts. Position them 2 inches in from the edge of the slab and where they will not interfere with doors or studs. This is very important. Once the concrete has set, there is no practical way you can move a bolt. The best you can do is hacksaw it flush with the slab's surface. Then, of course, you have to make do with one bolt less.

Bolts are positioned before the concrete is poured. Use a piece of stiff wire nailed to the form or a piece of wood with a hole drilled through it, and also nailed to the form to position the bolt (Fig. 159). Generally, you need 5 inches of bolt above the concrete. The block of wood with the hole drilled through it can also be used to position bolts that are mortared into openings in concrete block.

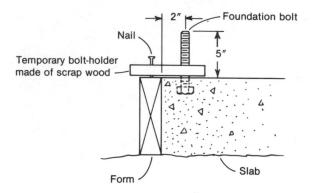

Fig. 159. *A piece of scrap lumber nailed to the form, with a hole used to position a foundation bolt while slab is poured.*

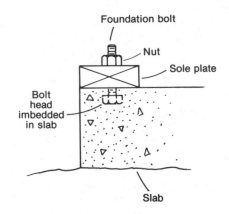

Fig. 160. *Sole plate, lowest member of a frame structure, held in place on slab by means of foundation bolt.*

CHAPTER THIRTEEN

Concrete Block Walls

The cutting and laying of concrete block was discussed in Chapter 4. This chapter covers several practical applications of those techniques. They are put together into one chapter because all concrete-block wall construction is essentially similar. The differences are only in their details.

FOUNDATION WALLS

Whether you are constructing a low foundation wall to carry a concrete-block garage or surround a slab foundation, or a full cellar beneath a thirty-room, wood-frame mansion, the procedure is exactly the same. Don't let the size of the job intimidate you. Simply follow the suggestions and adapt them to your specific project.

Locate the corners. Start by restretching the building lines from the batter boards (as you can see now, the batter boards have remained inviolate during excavation and the construction of the footing. With them in place, it is only a matter of minutes to relocate the proposed building). Drop the plumb bob from the points at which the lines cross. These crossover points are, of course, the building corner positions. Mark the position of the bob's point on the surface of the concrete footing. If the concrete is still soft, drive a nail into it. If not, simply scratch a deep mark. Now you have four nails or four marks that indicate the exact location of your building and its foundation. Note that the overall size of the block footing is exactly equal to the overall size of the frame of the building to be carried, without its sheathing. If a block building is to be erected on the block footing, overall footing dimensions will equal building dimensions.

Construct the corners. Now, your corner marks may fall so as to center the block on the footing, and then again they may not. Unless you have erred

Fig. 161. Mason is using a spirit level to make certain block is correctly positioned (*a plumb bob would have been easier*).

greatly in the construction and placement of the footing, and the block extends over or beyond the footing, ignore the footing in the placement of the corner block and all following blocks. Just make certain the corner of the corner block (you can use an end block or a stretcher block here) is in line with your corner mark or nail. If your footing is way off to one side, do not remove it, but construct an additional section of footing alongside.

Build each corner three or four courses high. Next, check the relative elevation of each corner. You want them to be less than ½ inch apart in height. If one corner is lower, mark that corner; when you lay up the following course, make that corner a bit higher. If you increase the joint thickness at the low corner a fraction of an inch at each course, you can easily correct elevation errors without the changes in joint thicknesses being at all visible.

Keep the corners vertical with the aid of your spirit level, as suggested in Chapter 4. In addition, tie your plumb bob to the crossover points on the building lines. Since these crossover points on the building lines remain fixed, and since this is exactly where you want your building corners to be, this method will provide the greatest accuracy. The point of the bob, of course, should just clear the outside corner of the building blocks.

If there are not going to be any fenestrations (windows, doors, vents, etc.) in the foundation wall, continue laying block till you get to the wall's top. If there are going to be fenestrations, review the following paragraphs on lintels before you do anything more than lay up the corners.

LINTELS

Doors, windows and the like are often fitted into fenestrations in foundations and other walls. Obviously, neither the doors nor the windows and their frames are anywhere near as strong as the portions of the wall they replace. Thus, to prevent the wall and its load above the fenestration from settling and breaking the door and/or window, a support must be provided. This support is called a lintel (Fig. 162). The lintel may be made of wood, steel or steel-reinforced concrete. Years ago, a large block of cut stone was often used.

Avoiding lintels. By proper design and fenestration placement, you can sometimes completely avoid the need for lintels. You don't need a lintel when the fenestration is the size of a single block or less. You do not need a lintel beneath a wood-frame building when there is no block above the fenestration and the opening is no more than 4 feet and in line with the floor joists overhead, or no more than 3 feet and across the line of floor joists (Fig. 164). The reason is that the building frame itself can span the above openings. You do not need a lintel above a metal doorframe when there is to be no more than two courses of block atop the frame and a wood-frame building above the block, and the doorframe is not more than 32 inches wide.

These are generally accepted figures, but individual building departments may be more limiting, so check before going further.

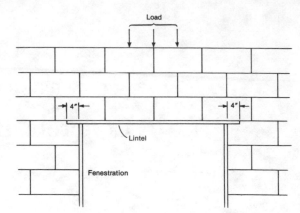

Fig. 162. A lintel is any type of reinforcement or support positioned over a wall fenestration to support a load placed on that fenestration. Note that a minimum of 4 inches of lintel must rest on the masonry at each side of the opening.

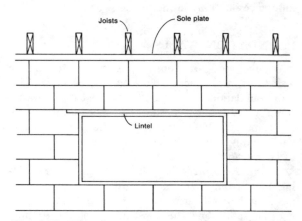

Fig. 163. This fenestration must have a lintel because there is nothing except the window frame to support the load.

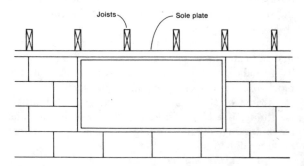

Fig. 164. This fenestration requires a lintel as it is more than 3 feet wide. If the joists ran in the other direction, fenestration could safely be 4 feet wide.

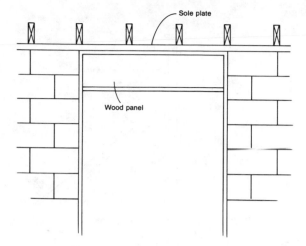

Fig. 165. Need for lintel over this door opening was eliminated by placing a wood panel over the door. (No block over the doorframe.)

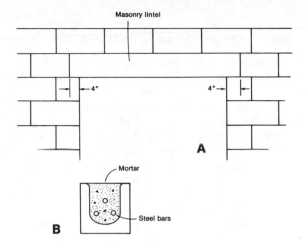

Fig. 167. Masonry lintel details: A. The lintel in place. B. End View. Size and number of steel bars are dictated by local building code.

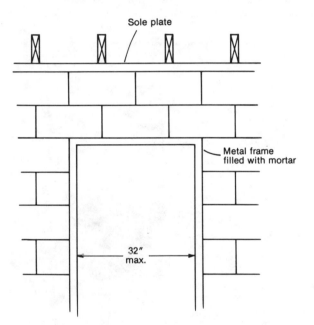

Fig. 166. No lintel is required here if door opening is 32 inches or less and frame is of metal. Hollow top of frame must be filled with mortar.

Installing a lintel. Build the wall ends adjoining or forming the sides of the opening for the door- or window frame until they are higher than the frame (or where the top of the frame will eventually be positioned). Then position the lintel over the opening. To install a reinforced concrete block lintel, select one longer than the width of the opening by at least 8 inches. This will permit 4 inches of the lintel

to rest on each block wall end. Lintel thickness should equal the thickness of the block that it is resting upon. Simply mortar the lintel in place just as you would a block, open-channel side up (Fig. 167). Place the reinforcing bars within the channel and mortar them in place, preferably a distance apart from one another.

To install a steel lintel, angle or T-shape, lay down a bed of mortar on the supporting masonry. Position the steel atop the mortar. Then cover the steel with a layer of mortar and position the block or brick on top of the mortar. As with the masonry lintel, there must be at least 4 inches of steel resting on the masonry supports.

A wood lintel is treated exactly like a masonry lintel, but a wood lintel should never be used in a basement where it can become wet and rot.

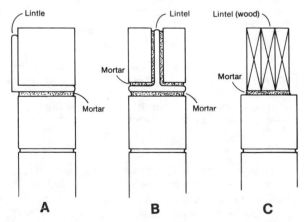

Fig. 168. Three types of lintels: A. Angled steel. B. T-shaped steel. C. Wood girder.

INSTALLING DOORS AND WINDOWS

Examine the building plans. See where the doors and windows, if any, are to be located within the foundation wall. See whether or not, by moving them a little, you may eliminate the need for extensive block cutting or even lintels.

Metal doorframe. Metal frames are always positioned so that their lower ends are covered by the concrete floor when it is poured. Metal doorframes rarely, if ever, are provided with sills.

Locate the position of the doorframe within the foundation wall which, of course, becomes the cellar or basement wall when the building is completed. With the aid of stakes and braces, raise and hold the frame in a vertical position (Fig. 169). If necessary, place stones or bricks beneath the frame ends. These stones and bricks will be covered by the concrete when it is poured. Check to make certain the frame was not warped or pulled out of shape, and that the door, when hung, will clear the finished concrete floor by ½ inch.

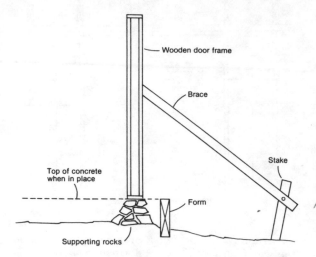

Fig. 170. How wood doorframe is positioned prior to pouring floor (on slab foundation door may be positioned following floor pouring).

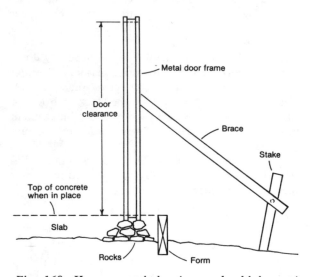

Fig. 169. How a metal doorframe should be positioned in relation to concrete floor about to be poured. Note that bottom ends of frame will be under the concrete, and that door clearance, after floor is poured, must be correct.

Wood doorframe. Construct a door buck of 2 × 8s with an outside dimension exactly equal to the outside dimension of the doorframe. Erect the door buck perfectly vertical and in line with the wall. (Fig. 170). Raise the door buck so that its lower

Fig. 171. Door buck in place after block wall has been erected. Wood triangles act as braces and are later removed if buck is left in place.

ends (and the doorframe to follow) will rest on top of the finished concrete and that the door's bottom will clear the concrete by ½ inch (more, if you are going to install a doorsill).

To save time and material, some masons do not use a buck but position the wood frame itself in line with the foundation wall they are erecting. Then,

(92)

Fig. 172. Window buck in place after wall has been erected.

Fig. 173. Window fenestration prior to finishing. Note use of end blocks at sides of window and sole plate at top of window opening.

when the wall is completed and the floor is in and the joints are fairly hard, they use concrete nails to nail the wood frame to the block. If you plan to use this method, do not use nails. Too often, the steel nail springs back and splits the wood frame. Instead, drill clearance holes in the frame and drill holes in the block. Then use lead anchors or similar masonry fasteners.

Some masons leave the buck permanently in the wall, then they either hang the door from the buck itself, or slip a smaller frame into the buck. The first arrangement is used on tool sheds and the like where appearance and a tight closure are unimportant. The second is used to produce a "different" sort of doorway.

Windows. You will find it much easier and faster to lay up the end blocks forming the sides of a window fenestration if you work with a window buck as a guide. As with a door buck, build the window buck from 2 × 8s. Make the outside buck width 1 or 2 inches wider than the frame to be installed, and 3 or so inches taller than the same frame.

If a lintel is necessary, position the buck so that its top will be flush with the tops of the nearest course of block. If there is no need for a lintel, position the buck so that its top will be flush with the top of the completed foundation wall, including capping, if

used (capping is discussed shortly).

To do the above you have to keep tabs on your progress (or later rip blocks out of the wall), and very likely raise the entire buck by setting it on pieces of wood and stone. When the mortar has set up and hardened, remove the buck.

Finishing doors and windows. In the case of a metal doorframe, you may want or need to fill openings between the frame and the block with mortar. In the case of a wood frame, you may want to cover the opening between the wood and the block with molding, or caulk it.

In the case of a window without an overhead lintel, wait until the building's sole plate is in place, then nail the top of the window frame to the sole plate. Make the frame perfectly vertical and temporarily lock it in place with wooden wedges. If the space between the side of the window and the block is small, all you need do is caulk it. If it is too large for caulking, it is filled and covered with mortar applied over the surface of the block at an angle. The space beneath the window frame is next covered with mortar applied at an angle (beveled) so that rainwater will run down and away from the window frame. Both the inner and outer sides of the frame may be treated this way.

Where there is a lintel, you can install the frame

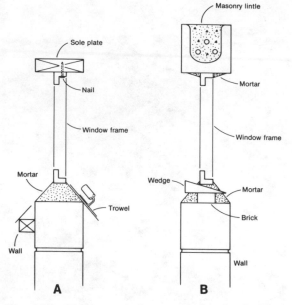

Fig. 174. How a window may be finished: A. Frame is nailed to sole plate, and mortar is beveled along bottom edge of frame. B. Frame is wedged in place, mortar is applied. Wedge is then removed, its space is filled with mortar.

Fig. 175. Finished window frame. Mortar has been applied in a neat bevel on three sides of the frame. Block has been parged and given a float finish.

as soon as you wish but, of course, you do not have the convenience of a sole plate to which you can fasten the frame. Instead, you have to wedge it carefully in place and apply the mortar as before. This is accomplished by applying just enough mortar at first to lock the frame in place. Then, when the mortar has set up, the wood wedges can be removed. Following, mortar is applied with a steel trowel. To secure a straight edge along the bottom of the window opening, have a helper hold a straight board against the side of the building to guide your steel trowel.

Foundation vents. Crawl spaces are always ventilated, otherwise accumulated moisture can damage the building. To save time and labor, secure vents with dimensions equal to the face dimensions of the block you are using. Then simply omit one block and use the vent in its place. A little mortar will secure it. Should you require a vent larger than the single-block-size vent, consider using two of them instead.

Capping a foundation wall. When a block foundation wall is to carry a block building, nothing is done to the top of the wall. The blocks that will form the sides of the building are simply laid atop the in-place block. When the block foundation wall is to carry a wood-frame structure, the wall is capped in any of several ways.

The simplest consists of stuffing rubblestones and even wet paper down the holes in the block, then filling the top 6 or 8 inches of block with mortar. Foundation bolts are then inserted into the wet mortar, heads down, and the mortar is struck off flush with the tops of the blocks. If a small board with a hole drilled through it is placed over each bolt, the board will hold the bolt upright until the mortar sets.

Most building departments will accept mortar fill as foundation wall capping. Others will not. Some require that you lay down 4-inch solids atop the wall. These are mortared in place. When the wall is made of 10-inch or thicker block, the solids will not cover it completely, as the solid is only 7⅝ inches wide. In such cases, the block is laid with its edge flush with the outside surface of the foundation wall. When and where the bolts can be positioned between block, there is no problem; the core hole is flushed full of mortar and the bolt head inserted as described. Where the bolt position falls in the center of a capping block, you have to split the block.

(94)

Fig. 176. Foundation wall has been capped by filling core holes in the block and positioning foundation bolts within the mortar.

Fig. 177. Parging a foundation wall. Mortar is applied with a smooth upstroke. Note hawk in mason's hand.

Waterproofing a foundation wall. This is accomplished in two stages. First the exterior of the foundation wall is covered with a solid, continuous layer of mortar. Then the mortar is covered with a layer of asphalt which reaches from the footing all the way up to the finished grade line (you do not want or need the black asphalt showing above the soil).

Start by removing all the loose mortar from the outside surface of the footing. Then knock off whatever mortar has adhered to the outside of the foundation wall (the wall will be clean if you have been cleaning up as you laid the block). Prepare a batch of mortar consisting of one part mortar cement and three parts sand. Add just sufficient water to make a thick cream. Then, with the aid of a hawk and a steel trowel, cover the exterior of the foundation wall with a ⅜- to ½-inch-thick layer of mortar. Place a glob of mortar on your hawk, then slide it onto the bottom of the trowel, which is held bottom side up. With a fluid backhand motion, apply the mortar to the block and spread it upward. The coat of mortar should begin atop the footing and curve upward. This is called parging or plastering.

Give the parging a good week or so to set hard. Then cover it with asphalt. Use the trowel type, the brush type is much too thin. Apply the asphalt with an old or inexpensive steel trowel. When done, you can discard the trowel.

FOUNDATION DRAINAGE

Dig a trench about 1 foot wide and 1 foot deep alongside the footing. Continue the trench past the house to some lower drainage point. (If you are on level ground and cannot lead the water away to a storm sewer or the like, the drainage trench will be useless.) Fill the trench with a couple of inches of crushed stone or gravel. Place a 4-inch bituminous or plastic drainpipe with holes in the sides atop the crushed stone. Raise the pipe where necessary to give it a pitch of ½ inch or more to the foot. You can do this with some stones. Next, cover the drainpipe with 6 inches or more of stone. Following, spread tar paper atop the stone and cover the paper with soil (Fig. 178). This arrangement permits water that may collect near the house footing to gather in the voids between the stones, enter the drainpipe and run off downhill away from the foundation. With the tar paper covered with several inches of soil, the excavation can now be backfilled,

(95)

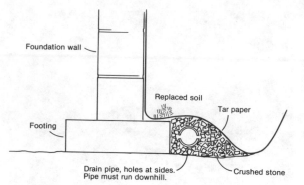

Fig. 178. Draining a foundation. Drainpipe must lead water away from building.

Fig. 179. Garden wall made of pierced block. Note that no block has been cut. Wall must be an even multiple of block dimensions. Courtesy National Concrete Masonry Association.

meaning the space alongside the foundation can be filled with soil. This portion of the job is complete.

GARDEN AND OTHER ORNAMENTAL WALLS

While garden and ornamental walls of concrete block do not support a load the way foundation walls do, they are still brittle in the sense that they cannot withstand tensile strains. They must be supported just as securely as foundation walls. If they are not, they will crack and possibly topple. In addition to a strong footing, the tops of block walls must be protected from the rain. Water collecting in the block and joints can freeze and break the wall.

Since garden walls are exposed, generally on both sides, a great deal more care must be taken when laying the block to make the joint widths even and regular, and to tool all the joints on both sides of the wall when it is completed. The alternative is to parge the wall after it's up. In this way you can hide inaccurate joints and minor errors.

Choice of block. Any type of block can be used for a garden wall. In addition to the standard blocks, there is a variety of ornamental block you can choose from. Some types have large, decorative openings to permit air to pass. Standard block can be cut as necessary; pierced ornamental block cannot. These blocks always have to be stacked, which means placed one directly atop another. This is not difficult to do, but stacking tends to magnify laying-up errors.

Avoiding block cutting. To avoid cutting block, which would result in ragged block ends showing, start by laying the end of the wall blocks as usual. Then, instead of building up the ends of the wall,

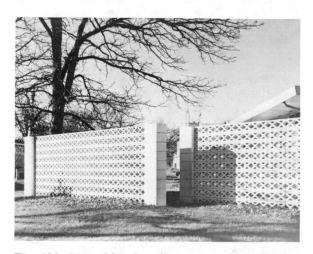

Fig. 180. Pierced-block wall terminating in pillars of stacked, double-ended end blocks. Courtesy National Concrete Masonry Association.

lay block across the length of the footing. When you come to the next to the last block, which would be the closure block for the first course, see how it fits. If it fits alongside the end block perfectly, fine: lay it in place. If it doesn't fit, move the end block.

Now, if you are stacking the block, you can build up the ends or lay successive courses right across, knowing that the block count will come out evenly. If you are alternating the courses, you can avoid

Fig. 181. "Slumped" concrete block used to make low garden wall. Color has been added to mortar to make joints more visible. Courtesy National Concrete Masonry Association.

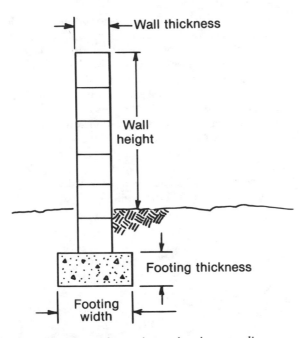

Fig. 182. Major dimensions of a free-standing concrete block wall.

block cutting by using half-length blocks where you need them, instead of cutting block.

Wall height. The accompanying table gives maximum above-grade heights for block walls of varying block sizes. If you plan or need to go higher, it is advisable to secure local engineering advice. In some localities you will need a permit to build a garden wall higher than herewith recommended. The building department will then advise on the changes necessary to make a higher wall safe.

MINIMUM RECOMMENDED DIMENSIONS FOR
FREE-STANDING CONCRETE BLOCK WALLS

Wall height	Wall thickness	Footing width	Footing thickness
3 feet	6 inches	10 inches	4 inches
4	6	12	5
5	10	18	7
6	12	20	8

Capping garden walls. Any of several methods may be used for keeping rainwater from entering the wall. The simplest is merely to lay down a 2-inch-thick layer of mortar atop the wall and then round it off or angle the top surface of the mortar.

Another method of capping a garden wall utilizes vitrified tile coping. This coping comes in a number of sizes and shapes. The top of the wall is covered with a bed of mortar. The coping is placed atop the wall so that the end of one piece of coping overlaps the other.

Flagstone and slate are also used to cap block walls. Either stone may be cut to a width slightly greater than the thickness of the wall, and positioned lengthwise on mortar atop the wall. When this is done, the surface of the mortar is angled so that the stones tilt slightly to one side to provide water runoff. An alternate method that is possibly more at-

Fig. 183. A block wall may be topped with mortar shaped into a curved surface with a trowel.

(97)

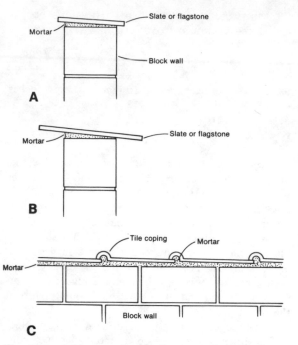

Fig. 184. A block wall may be topped with narrow slate or flagstone (A) wide slate or flagstone laid at an obvious angle (B) or with tile coping laid up in mortar as shown (C).

tractive consists of placing the stones lengthwise across the top of the wall so that the ends of the stones project beyond the sides of the wall by a few inches. When this is done, the surface of the supporting mortar is angled even more so that the resultant pitch of the stone capping is quite obvious (Fig. 184).

CONCRETE BLOCK RETAINING WALLS

Whereas a supporting wall of concrete block is easily several times stronger than any building you or I may place on top of it, a retaining wall is comparatively weak in relation to the side pressure that will be exerted by the hill it retains. This pressure develops as soon as rain falls and the earth behind the wall becomes soft and slumps downward against the concrete block. As a result, concrete block retaining walls must be constructed of large size block, reinforced one way or another depending on wall height, and never should be more than 5 feet high without engineering assistance. Most building codes demand a "designed" retaining wall, meaning that it be designed by a licensed engineer. There is, of

course, good reason for this. A 5-foot or higher wall that topples under the onslaught of mud and water can be deadly destructive.

General construction. A block retaining wall requires the same footing and footing placement as a foundation wall or a garden wall. The blocks are laid up exactly the same as for any other block wall. The top of the wall must be capped, again using any suitable method. The only difference between a block retaining wall and any of the other types of walls discussed so far is drainage.

Drainage. Water that collects behind a block wall can freeze, and in doing so exert much more pressure than the soil alone. In many cases, it is frost that knocks the wall down and not the weight of the retained hillside.

Several methods can be used to drain the wall (Fig. 185). One method consists of simply leaving 1- or 2-inch spaces between every fourth block. Another consists of positioning a short length of ceramic or plastic pipe 5 inches in diameter in the wall. The pipe is held in place with mortar. Its front end is positioned flush with the surface of the wall. The pipe or the openings are positioned as low as is practical in the face of the wall. If the wall is more than 3 feet high, install a second row of drains. Place a number of stones behind all the openings. The stones act as sieves and retain the soil while permitting the water to flow through.

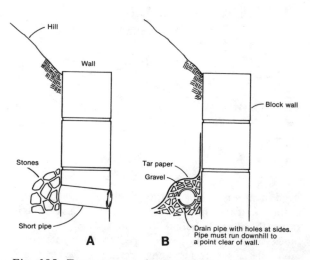

Fig. 185. Two ways to drain a block retaining wall. A short pipe at an angle through the wall. (A) or one or more long drainpipes at an angle laid parallel to the wall and reaching beyond the end or ends of the wall (B).

Another method of draining a retaining wall that does not require holes through the wall consists of transverse pipes. The wall is erected, but the space behind the wall is not filled with soil as yet. Instead, several inches of crushed stone or gravel are spread over the bottom of this space. Next, two lengths of perforated plastic or bituminous drainpipe are positioned atop the stone, end to end. Pipe lengths should be such that their ends extend 8 or so inches beyond the ends of the wall. The point where the two pipes meet is raised a few inches with the help of stones. Both pipes are covered with 6 or more inches of stone. Next, the stone is covered with tar paper and then sufficient soil to firmly fill the space behind the wall.

When water collects behind the wall, the water enters the pipes, and since they are pitched toward the ends of the wall, the water is let out from behind the retaining wall.

Reinforcing a block wall. Where space permits, the retaining wall can be reinforced by buttresses. These are short walls constructed at right angles to the main wall. They require footings similar to the main wall.

Block retaining walls can also be reinforced by inserting rebars (steel reinforcing bars) within the block and footing. You can do this by forcing the steel bars through the footing while the concrete is still soft. The bars, spaced a measured distance apart, are held in a vertical position until the concrete hardens. The distance between bars depends on the number of bars you want to use and how the holes in the blocks work out. Generally, you have to stack the block if you go up more than two or three courses. In any case, the number of bars and their thickness and spacing will depend on the height of the wall and the nature of the hill behind it. All this requires local engineering assistance.

RIP RAP

To keep a hillside that won't support grass or other vegetation from washing down with each rain, it may be covered with rip rap. This is an engineering term for covering the hillside with flat stones or concrete block. If you choose block, use 6-inch solids. Make the hillside as smooth as you can, then lay the blocks on their backs, side by side. Use as many blocks as you need to cover the hill. The weight of the blocks keeps them in place. Nothing more need be done.

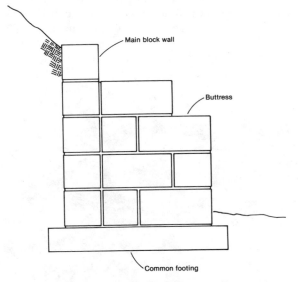

Fig. 186. *How a short wall is buttressed up against a block retaining wall to strengthen it.*

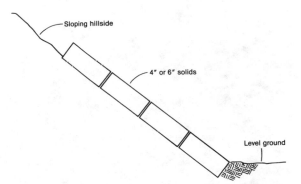

Fig. 187. *Rip rap: block laid flat atop a hillside to hold the soil in place.*

CHAPTER FOURTEEN
Stone Walls

Be it ever so humble, a home encircled by a stone wall is always more charming, more attractive than the same structure alone. The why of it is difficult to explain, but it is so. A fence or a wall edges a property, says this is it, this is where it begins and ends. A wall provides the finishing touch. When the boundary line is defined in stone, the estate, be it large or small, appears stable, permanent, a home one can rely upon for centuries. That, of course, is the nature of stone. When a stone wall is properly made it will last for centuries, even for a millennium; many have.

TYPES OF STONE WALLS

If we divide all types of stone walls into basic categories, we find there are only two: dry and mortared. A dry stone wall is a wall put up without the benefit and aid of mortar. One stone is simply laid atop another. A mortared stone wall consists of stones joined by mortar.

Made without mortar, the dry stone wall is flexible. It is not adversely affected by frost heave. It does not require a foundation resting on below-the-frost-line subsoil. The dry wall can be laid right

Fig. 188. Front of a dry stone wall. Note how stones have been selected and positioned to overlap one another.

Fig. 189. End view of same wall. Note how overlapping has been carried through and that most stones lie horizontally.

down on the grass. Since the dry wall depends on the weight of the stones alone to hold it together, the size and shape of the stones are important, as are their placement. In effect, the size and shape of the stones more or less dictate the height, width and cross-sectional shape of the wall. The shape and size of the stones also affect the ease with which the wall can be constructed. Large, flat, straight-sided stones are easily laid up into a wall with vertical or nearly vertical sides. Round, cannonball rocks cannot be assembled to form a wall with vertical sides. Without mortar, the best you can do with cannonballs is to build a wall with a pyramidal cross-section. Odd-shaped stones require lots of trying and fitting in order to position them correctly. Small stones—stones less than a foot long and across—are unstable when piled atop one another to form a wall.

The mortared stone wall, when completed, is a single, monolithic lump of stone. It is not in the least flexible. It must rest on a firm below-the-frost-line foundation. If it doesn't, it will crack. Just how badly it may crack depends on the size of the wall, the nature of the soil on which it rests and the extent of the frost. Cracking is not something that can be predicted with accuracy. However, the larger and higher the wall, the softer and wetter the earth, and the greater the frost, the greater the cracking. In really bad cases, frost heave can knock the wall down.

BUILDING DRY STONE WALLS

Choice of stone. Since the size and shape of the stones that will be used to construct the wall have a tremendous effect on the height and cross-sectional shape of the wall, the wise mason determines the stone he is going to use before he plans his wall.

The type of stones you will use (i.e., the size and shape) depends on chance and your willingness to work or spend money. If you build with what is available for the taking, stone that you find on your property or your neighbor's, or haul home from construction sites (builders are usually only too glad to have stone carted away), you are most likely to have a mixed bag. The stones will be of all sizes and shapes, not to mention color and texture. If you wish to expend the effort in good clean exercise, you can cut and shape the stones. Stone cutting and the required tools are discussed in Chapter 5.

If you would rather spend money than energy, you can purchase stone cut to your needs and desires. But bear in mind that the terms quarry stone simply means the stone was cut from a quarry. It doesn't mean each stone is nice and flat with four square corners. Each quarry has its own termi-

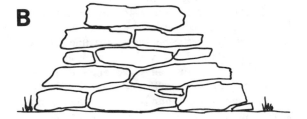

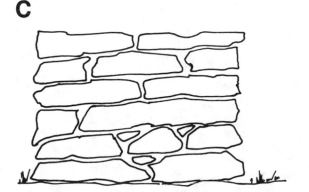

Fig. 190. How shapes of stones more or less dictate base width-to-height of stone wall: A. Round stones cannot be piled in anything other than a pyramidal shape. B. Semiround stones can be piled somewhat higher. C. Flat or nearly flat stones can be piled to form a wall with vertical or nearly vertical sides.

nology, but a truckload of loose stones from any quarry is one-third air. For example, if you order 6 cubic yards of quarry run, their power shovel will fill the back of a 6-yard dump truck with stone. Since the stones cannot possibly pack together tightly, roughly one third of the space is air and of the 6 yards of stone, for which you pay the going rate for 6 yards, only approximately 4 yards will be solid rock.

Layout. When you know the general size and shape of the stones you are going to use, check your stone shape and desired wall height against the accompanying table. The table will give you the recommended minimum base width for a given wall height and stone shape. If you do not make your wall base as wide as recommended, the wall will not be stable.

In fact, if you try to erect a nearly vertical wall with cannonball boulders, the stones will not remain in place long enough for you to turn away.

Knowing the base width of your planned wall, lay it out in the position you wish to build it. Use either long boards placed on the earth or lines stretched from pegs. Space the boards or the lines just as far apart as the desired base width of the wall.

Construction. You can leave grass in place and build your dry stone wall directly on top. Weeds and the like should be removed. They will continue to grow and may come up right between the stones, which will make their removal very difficult.

Ignore ground slope if the slope is in line with the wall. In other words, your wall can run up and down hillsides without any alteration in its construction. However, there can be no ground slope or pitch from one side of the wall to the other. When this is the case, or when and where there may be a little mound on the surface of the earth, excavate as necessary. The earth does not need to be perfectly level, just level to the eye.

Disperse your stones alongside one or both sides of the lines marking the wall, leaving two feet of clearance between the wall lines and the stones you are going to use to build the wall. Doing this makes all the stones readily visible and accessible. In this way you can easily examine all the stones when you are seeking a stone of a certain size and shape.

If your wall is to be low and narrow and your stones are large, you may be able to construct the wall just one stone wide. In which case, position the first course or row of stones within the guidelines. If all the stones are flat or nearly flat, fine. Go on to the second course. If some of the stones are not flat, either remove some soil from beneath their high ends or raise their low ends with small stones and

WALL HEIGHT VS WIDTH VS STONE SHAPE VS MORTAR

DRY WALL WIDTH

Height	Round stones	Flat and oval stones	Flat stones
2 feet	3 feet	2.6 feet	1.5 feet
3	4.5	3.9	2
4	6	5.2	2.7
5	7	6.5	3.5

SEMI-DRY WALL WIDTH

Height	All stone shapes
2 feet	1.5 feet
3	2
4	2.5
5	3

FULLY MORTARED WALL WIDTH

Height (above ground)	All stone shapes
2 feet	1 feet
3	1.5
4	1.75
5	2

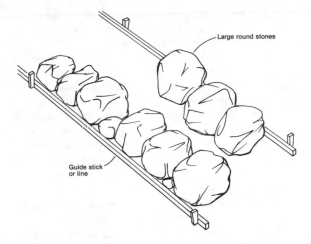

Fig. 191. *Use your roundest, largest stones to form the bottom courses of your wall. A pair of lines or sticks are all you need to guide you.*

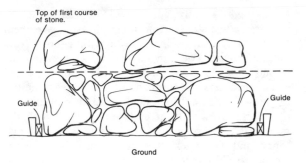

Fig. 192. *Make the top of the first course of stones fairly level by filling the spaces between them with small stones.*

stone chips. You need the surface of the first courses and all subsequent courses to be reasonably flat and level, or as flat and level as is practical. Start the second course with a half-size stone. Follow with a full-size stone, and so on. When the stones are flat they are laid up exactly as is concrete block.

Assume you are building a wall two or more stones wide with flat stones. Proceed as before and lay half of the first course for a distance of a few yards. Then start the second half of the first course with half a stone, followed by a full-size stone. Level the stones as suggested. Construct the second course of full- and half-size stones. Position the full-size stones so they lie across the width of the wall and rest on two stones. Where necessary, use small stones and chips to make the surface of the second course of stones reasonably level. The third course will consist of two rows of stones running the length of the wall. Thus, each course or layer of stones will overlap two or more stones, and the weight of the stones themselves will lock the wall into one strong mass.

If you are working with a mixed bag of stones, which is much more likely, start by placing the largest and roundest stones on the ground, within the lines. Turn the stones so that their flat sides—if they have any—are directed outward and up. If there is only one flat or nearly flat side, it must be turned up. If a stone seems likely to turn because its bottom is uneven or round, either excavate a little hole beneath the round projection or stuff small stones alongside so that it cannot turn.

Next, fill the spaces between the first course of rocks with small stones and stone chips. Where possible, try to find stones that will fit into the openings. In both these operations, work toward a relatively level surface, flush with the top surfaces of the large stones (Fig. 192). If necessary, cut stones to better fit the cracks and crevices. Bear in mind that you need the level supporting surface so that you can lay down the second course of stones. This done, you can lay the second course. And, as previously described, each stone in the second course must rest on two or more stones. Spaces between the second course of stones are filled with small stones as before.

When you have large flat stones with square edges, you can build your wall straight up. When you have round stones or a mixed bag of stones, you are going to have to make each succeeding course a bit smaller and narrower than the preceding course. When are working with loose, unselected quarry stone, you are going to have a lot of triangular-shaped and other oddly shaped stones on hand. Just keep twisting and rotating these stones until they fit into place. Think of it as a giant jigsaw puzzle.

Topping the wall. You can simply quit when you

Fig. 193. *If you simply pile your stones one atop the other as shown, you wall will fall down in short order.*

are as high as you want to be. You can select the largest, flattest stones from your pile and save them for the top; their weight prevents little boys from toppling them. You can fill the spaces between the topmost stones with mortar, letting some of the stone surfaces show through. You can cover the top of the wall with a 2-inch-thick layer of mortar, with the mortar troweled flat and smooth. Although the mortared top results in a nonflexible length of masonry, and will crack, the cracks occur between the stones or a distance apart. The result is a series of large sections of masonry topping the wall and doesn't look bad.

Mortar for stonework. You can, if you wish, use the same mortar mix suggested for working with block and brick. That is a 1:2 or 1:3 mix of mortar cement and sand. Most masons, however, use regular cement (which has no lime) when they work with stone. The usual mix ratio is 1:3. Professional masons rarely wash stone before they apply mortar. This is an error. When you want a good strong joint, you must wash the stone free of all mud and muck. Unless quarried stone has been resting on mud, it does not need to be washed.

SEMI-DRY WALLS

The semi-dry wall is a practical compromise between a wall that is inflexible and one that is flexible. It is also a practical answer to the problem of building a vertically sided wall with poorly shaped stones. The semi-dry wall will crack when heaved by frost, but the cracks are invisible. Since the mortar permanently joins the small stones into large stones and in some places locks the entire wall together, its height-to-width ratio can be a little more than a dry wall made of flat stones (Fig. 194).

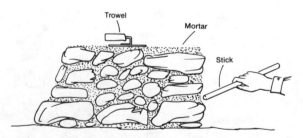

Fig. 194. How a fully mortared wall may be constructed of small, odd-shaped stones. The wall is topped by a layer of mortar. A partially mortared wall would simply include less mortar.

Construction. Begin as before. Stretch the lines, and position your largest, roundest stones on the ground. But before filling the spaces with small stones, partially fill the spaces with mortar. Use the suggested 1:3 mix, but only enough water to enable you to form the mortar into a ball—no more. Now force the stones into the in-place mortar. Using mortar and stones, construct the level platform on which you will place the second course of stones. If a second-course stone is properly shaped for its position, fine. Just lay it down. If its shape is wrong, if it needs to be propped up by smaller stones, use a little mortar to lock these stones in place. In other words, use mortar and small stones to cement the big stones in place. But take care to keep the mortar a distance in from the sides of the wall. Doing so will keep the mortar relatively invisible and the wall will look dry. If some mortar is forced into view by the weight of the stones, use a pointed stick to force it back out of sight.

The semi-dry wall can be topped by any of the methods previously discussed.

FULLY MORTARED WALL

The advantage of joining all the stones together with mortar is that it can be a wall with a much greater height-to-width ratio than you could make with just a little mortar or no mortar at all. The fully mortared wall cannot be taken apart by little boys or entered by small animals, but it is otherwise no more permanent than the other two types of walls. The disadvantage of the mortared wall is that it requires a footing to protect it against frost heave.

Footing dimensions. Depending on wall height, thickness and ground conditions, footing requirements run from zero to a poured slab twice as wide as the wall's thickness, and as high as the wall's thickness. In other words, approximately the same rule of thumb that applies to a block-wall footing is used for calculating a mortared stone-wall footing.

Construction. Excavate the necessary trench to a depth below the frost line. If there is no frost line, excavate below the topsoil to the firm subsoil. Construct a footing in any of the ways described in Chapter 11. You can pour the concrete into the bottom of the trench or build a form within the trench and pour it full of concrete.

Recommended Minimum Dimensions Mortared Stone-Wall Footings on Firm Subsoil

Wall height above ground	Wall width	Footing width	Footing thickness
2 feet	1 feet	1 feet	8 inches
3	1.5	1.5	8
4	1.75	2.	8
5	1.75	2.5	10

Stretch two lines down the length of the footing. Position the lines about 1 foot above the footing and just as far apart as the desired width of the wall. Spread a 1-inch-thick layer of mortar atop the footing. Make the layer of mortar 1 or 2 feet long and just as wide as the desired width of the wall. Place your stones atop the mortar. Spread more mortar and lay down more stones until you have a first course of stones 5 or more feet long. As described previously, fill the spaces between the in-place stones with mortar and more stones to produce the desired level surface. Spread a second layer of mortar atop this surface and position your second course of stones. Fill in the spaces. Continue doing so until your following course will be aboveground and visible. From here on up, proceed as before, but take care to place the best side of the stone facing outward. This procedure is repeated until the entire wall is completed. The wall may be topped by any of the methods previously described except, of course, dry stone alone.

Working with cut stone. Cut stone, by masonry definition, is stone cut into smooth, orthogonal shapes. When using such stones, the beneath-earth layers or courses are laid down as previously suggested, that is to say, with little concern for outward appearances. The visible courses are laid up with great care. The stones are matched to make certain they all fit their allotted spaces, that the top course comes to a level surface or line and that all stones overlap one another. Joints are tooled just as are block and brick joints, and with the same tools.

STONE RETAINING WALLS

In construction, a stone retaining wall differs little from a similar garden or ornamental wall. The difference in load, however, is usually tremendous. Whereas a garden wall has little but its own weight

Fig. 195. *Here you can see the large quantity of mortar that has been placed between the stones. Doing so eliminates the need for a lot of cutting and makes almost all stones useful.*

Fig. 196. *Spaces between top course of stones have been filled with mortar. Mortar is now brushed to give it a sandy appearance.*

Fig. 197. *Completed portion of wall. Note how wide the mortar joints are. This is the easiest type of mortared wall to make.*

to support, a retaining wall may be called upon to sustain a side pressure of many tons. That is why most municipalities require that retaining walls higher than 5 feet be designed by a licensed engineer.

The side pressure on a retaining wall depends on the nature of the soil behind the wall and its height. A level stretch of land no higher than the wall itself places a minimum load on the wall. A hill, meaning the surface of the earth behind the wall angles upward, places a maximum load on the wall. If the earth is composed mainly of rock and sand, the load will be relatively small because rocks and sand tend to remain in place. If the soil is mainly clay, loading will be minimal in dry weather; given sufficient water, however, the clay will turn into mud and the hill will collapse sideways. Thus, it is imperative that water is not permitted to collect in any kind of soil behind the wall. Not only will water liquefy clay, but should the water freeze, it can exert as much pressure against the wall as a hill of mud.

Design choice. We can build the retaining wall perfectly vertical, just as we can build a garden wall of stone, or we can lean or slope the retaining wall up against the earth. In doing so, the potential load on the wall is greatly reduced. In either design, vertical or sloped, we can lay the stones up dry, partially mortared or fully mortared. Laid up dry, the wall will drain itself; laid up partially or fully mortared, water will not pass easily through the wall and the wall must be drained. Any of the methods suggested in Chapter 13 for draining concrete-block walls can be used with stone wall. And, like the garden wall of stone, the retaining wall can be capped by any of the methods previously mentioned. In addition, the retaining wall can be capped or topped by sod. Sod placed atop the retaining wall will grow and remain green, as it will be watered by rain and whatever water comes down the hillside.

Constructing a dry, vertical stone retaining wall. With a shovel and pick, cut a flat-bottomed, horizontal shelf into the hill or terrace far enough to permit you to lay all the stones on firm, level soil. Lay up the stones as suggested previously for constructing a dry stone wall. In other words, you will erect a dry stone wall alongside the hill or terrace. When you have completed the wall, fill the space behind it with stones. Position the larger stones at the bottom, the smaller stones on top, and fill in the remaining space with soil. Top the wall as suggested.

RECOMMENDED MINIMUM WIDTHS OF VARIOUS STONE RETAINING WALLS

Dry and semi-dry, retaining a terrace*

Total height**	Vertical wall width	30° or greater wall slope, wall width
2 feet	2 feet	1.5 feet
3	2.5	2
4	3	2.5
5	3.5	3

Dry and semi-dry, retaining a hill*

2 feet	2 feet	1 feet
3	3	1.5
4	3.5	2.5
5	4	3

Fully mortared wall, retaining a terrace

2 feet	2 feet	1.5 feet
3	2.5	2
4	3	2.25
5	3.5	3

Fully mortared wall, retaining a hill

2 feet	2 feet	2 feet
3	2.5	2.5
4	3	2.75
5	4	3.5

* All stones laid up dry in a vertical wall are flat.
** Height is measured from top of footing, if used, or bottom of wall.

Constructing a sloped, dry stone retaining wall. Since a portion of the wall's weight rests against the earth, you do not need perfectly flat stones for this type of wall. You can use some oval and odd-shaped stones, but not too many or the wall will be weak.

Proceed as before. Cut a shelf into the hill or terrace. Make the vertical portion of the shelf (the side of the hill or terrace) angle backward. Lay the stones up as suggested, but fill the space behind the stones with smaller stones and soil as you go. In other words, back-fill as you build your wall. Position each successive course a little back of the course upon which the stones rest. In this way you will slope the wall and still keep each stone more or less on a horizontal plane (Fig. 198).

An alternate method, supposedly originated by the Romans, begins the same way. The first course is positioned. The space between the stones and the earth to the rear of the wall is filled with small stones. Then the first course of stones is covered

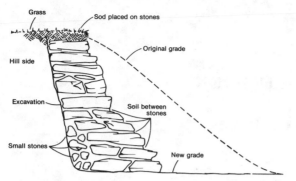

Fig. 198. Sloped, dry stone retaining wall made by covering courses of stones with soil instead of mortar. Top of wall is covered with sod.

Fig. 199. An example of a sloped, dry stone retaining wall made according to the Roman method.

with an inch or two of soil. The surface of this layer of earth is made smooth with a shovel or trowel, and the second course of stones is positioned atop the soil. This is continued until the wall is completed.

Constructing a semi-dry stone wall, vertical or sloped. Proceed as previously suggested. Cut the shelf into the earth and lay up the stones with a little mortar. If the wall is to be vertical, build vertically. If the wall is to slope, offset each course a few inches rearward as you go up. But each stone or each large stone is always more or less on a horizontal plane.

Unlike the dry wall, the semi-dry retaining wall must be drained. One simple method is to place lengths of 2 × 3s transverse to the width of the wall as you go. Then, before the mortar has set up, pull the pieces of wood out of the wall. Doing so will leave drain holes.

Constructing a fully mortared stone retaining wall, vertical or sloped. Proceed as before. Cut the shelf into the earth. Then dig a trench and install a footing exactly the same as specified for supporting a mortared garden wall. Build your mortared retaining wall on the footing. The wall does not have to be centered on its footing; it can be closer to the hill or terrace it will restrain than to the center of the footing.

The fully mortared retaining wall must be drained. Use any of the methods suggested in Chapter 13 for draining concrete block retaining walls.

RIP RAP

The side of a hill can be covered with stones laid down on their flat sides to keep the earth from running away and to discourage vegetation. As when concrete blocks are laid this way, stones are also called rip rap. Simply begin at the bottom of the hillside and lay the stones down, one adjoining another.

CHAPTER FIFTEEN

Cellar Floors

While a concrete cellar floor is simply a slab of concrete lying in a cellar—just flat work, as a mason would describe it—the making of a concrete cellar floor differs in a number of ways from the making of patios and the like, and is probably the most physically difficult job in modern masonry. The pour is confined by the walls of the cellar, which means you cannot get to one side to screed and finish. Most of the time you have to screed by pulling the screed toward yourself, and finish working on your knees. When the concrete can only be brought into the cellar from a single window or doorway, the concrete has to be moved by hand across the floor. Then you have to muck through the slush to screed it.

Therefore, do not attempt to do a cellar by yourself or even with a single helper. At a minimum, three people are needed: two on the screed while the third works the rake or shovel to assist screeding.

All crew members must have knee-high rubber boots because they will be walking through soft concrete, and all should have an iron rake and shovel, plus the usual tools for finishing concrete.

Timing. Cellar floors can be laid anytime the foundation walls are up. Professionals usually wait until the building is roughed out, meaning the house frame and roof are up. The reason for waiting is that the roof eliminates the possibility of a heavy downpour ruining their work. And, even if the concrete has set up and is unharmed, a cellar full of water is a nuisance and must be pumped dry.

However, if it is midsummer and there is little or no chance of rain, the benefits derived from working without a roof are worth chancing rain damage: you will have plenty of light; you can pour the slab from all sides so that little concrete movement by hand is necessary; drying and setting up are speeded so that you can use a slightly wetter mix and still not wait hours for it to be workable.

Cellar preparation. Make certain that all the pipes and electrical cables that must enter the building through the cellar floor are in place. If any of the pipes are merely sitting there and can be moved out of position when you work, lock them in place with a mound or two of concrete or mortar.

Clear the cellar of all wood chips, loose stones and loose soil. Clean the top of the footing and the bottom 6 inches of foundation wall. If there is mud on these surfaces, remove it with water and a steel scrub brush. If there are mortar splatters, remove them with a cold chisel and hammer. You want the top of the footing and the bottom of the foundation wall to be fairly smooth and clean. Dirt will prevent proper adhesion between the concrete and the foundation. Roughness will interfere with screeding.

Preparing the base. Since you have dug several feet into the earth for your cellar, you are now below the frost line, so frost is no problem. You can therefore pour the floor directly on the earth, or you can provide a base of crushed stone or gravel.

If the bottom of your footings rest on the bottom of the cellar excavation, then the footing stands 8 or 12 inches higher than the soil. Since you want your concrete floor to rest on top of your footing, this condition calls for 12 to 16 inches of concrete overall. To save on concrete, cover the soil with sufficient crushed stone or gravel to equal the height of the footing. This done, you will have a level sur-

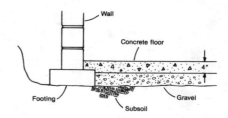

Fig. 200. Cross-sectional view of concrete floor atop a gravel base. Note that slab reaches to foundation wall.

face from foundation wall to foundation wall (Fig. 200).

If, on the other hand, you have dug trenches in the bottom of your cellar excavation and have placed the footings within the trenches, then the top of the footing will be in line with the surface of the cellar floor soil. In such cases, you will need no stone atop the soil, or at least considerably less.

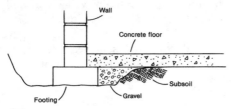

Fig. 201. Concrete floor atop subsoil. Space near footing filled with gravel.

Installing the guides. Use 2 × 4s on edge for guides. Position them in line with the direction from which you will receive the concrete. In other words, you want to move the concrete between the guides, not across them. Use a minimum of three guides. Position two guides a foot or two from the walls; more or less center the third between them. If the space between any two guides is much more than 8 feet, use four guides. Support the guides on short stakes driven into the earth. The tops of the stakes must be flush to or below the top surfaces of the guides. Raise or lower the guides until all their surfaces are on a single plane and that plane is 4 inches above the top of the footing. Use a spirit level and a straight-edged board laid across two or more guides to help you do this. Place stones or globs of concrete or mortar beneath the guides to prevent them from descending when you place the screed on top (Fig. 202).

Moisture barrier. To reduce the possibility that water may seep through the cellar slab and into the cellar, a moisture barrier may be placed beneath the slab (Fig. 203). The barrier is not a guarantee that no water will enter this way; it is only a barrier. Generally the barrier, which is a sheet of plastic, is laid on top of crushed stone or gravel, which has been made level with the top of the footing. In addition, if crushed stone is laid down, the stones should be tamped with a metal tamper to make certain no stone points project upward. If some do, they may make holes in the plastic when you and your crew walk on it.

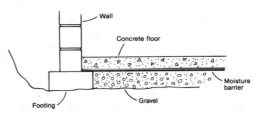

Fig. 203. Concrete floor atop moisture barrier atop gravel base.

The plastic sheet or sheets are stretched from wall to wall. The ends of the sheets rest atop the footing. Where necessary, sheets are overlapped by a foot or two to make a sealed joint.

Naturally, with the moisture barrier in place, you cannot use stakes to support the guides. Stakes would pierce the barrier and render it useless. The solution lies in placing 4 × 4s atop the barrier in the same position you would put the stake-supported guides (Fig. 204). Since the 4 × 4s are not nailed to stakes, they will rise and float atop the fresh concrete. To prevent this, cover each 4 × 4 with several shovelfuls of concrete, spaced a distance apart. The globs of concrete will hold the guides down. When you screed, the screed will remove the concrete from the tops of the guides.

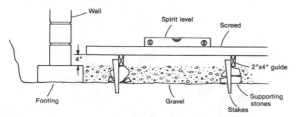

Fig. 202. How stakes are used to support guides for concrete cellar floor. Stones keep guides from possibly being pushed into the earth.

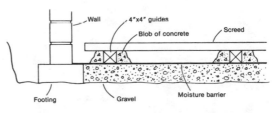

Fig. 204. How 4 × 4s are used as guides atop a moisture barrier. Blobs of concrete hold guides in place.

Pouring and screeding. The easiest mix to use is the high-sand formula (1:2¼:3) and ½-inch gravel or very small crushed stones. If you are pouring into an open cellar and can reach most of the floor with the truck's chute, use a normal quantity of water in the mix. If you are forced to move the concrete a goodly distance across the floor because you cannot move the chute, use a larger quantity of water to make a "wet" mix. The more water, the easier the concrete is to move along with rake and shovel.

In any case, you must pour in line with your guides. You cannot push concrete over the guides. If you do, you surely will move them even if the guides are nailed to stakes. And you must screed toward the source of the concrete, the chute. You don't want to screed yourself into a corner like a painter who starts at the single door of a room and paints backward until he finds himself in a corner, locked in by a painted floor.

If you have divided the floor into three sections, pour and screed the two side sections first, then the center section. This is the easiest way. If you permit the side sections to set up a little, you won't sink into them very much when you pour and screed the center section.

As soon as the concrete has almost set up and you can walk on it without sinking in more than a fraction of an inch, remove the guides. Then fill the channels left by the guides with fresh concrete. Using a straight stick, screed the concrete you just placed in the channels. When doing so, let the adjoining surfaces of the concrete act as your guides.

Finishing. Now you have to float the floor. Start with the bull float at the far end of the floor and work your way back toward the door. Follow up with a darby, then follow that with a hand float. This is next to the final finishing step.

The surface of the concrete is still workable, and obviously you do not want to leave footprints. So, you walk to the far end of the floor with the aid of a pair of flat boards which you lay down and pick up as you go. Then you kneel down on the boards and hand-float the concrete. As you work your way back toward the door, you move the boards one at a time so that you never step on the fresh concrete.

A float finish is fine for a garage floor or a floor on which you are going to lay tile. But if you plan to paint the floor, you have to steel-trowel the entire surface after you float it. A float finish will never take paint properly; it will always have a sandy texture. A float finish in a busy basement or cellar also produces dust. The sandy texture is abrasive and tends to grind whatever is moved over its surface. The sandy texture also holds dust, which makes it difficult to clean and keep clean.

Fig. 205. To keep from marring fresh concrete, mason works from a pair of boards. He has a float in one hand and a steel trowel in the other. In this way, he doesn't have to crawl over the surface of the concrete twice.

CHAPTER SIXTEEN

Brick Walls

Until the early 1900s, concrete block was not commercially available. For unit masonry construction, masons had a choice of stone or brick, and brick was by far the cheapest in areas where clay and coal were available. As a result, well into the 1920s most homes and commercial buildings in the European-American world not constructed of wood were constructed of brick. Today, with the large variety of block available and the greater amount of labor required to lay brick, brick is rarely used for building foundations or walls. But brick has a charm and beauty concrete block cannot duplicate, so brick is still used in large quantities for decorative purposes. Thus, you will find brick garden walls, brick room dividers, brick planters, brick-veneered walls, etc., indoors and out.

None of the brickwork is particularly difficult. In many ways, brickwork is easier than working with block. The very small size and weight of a brick in comparison to the average block makes it so. The smaller comparative size of brick in relation to block also makes brick a bit more tolerant of lay-up inaccuracies than block. It is easier to see that an 18-inch-long block is laid up at an angle than it is to see the same error in an 8-inch block. It is also possible to recess brick joints so that they are somewhat out of sight and unnoticed. This is not usually done with block.

FOOTINGS

When a brick wall is completed and its mortar has set up and hardened, it is a solid, monolithic piece of masonry. It is not flexible to any degree, and its tensile strength is comparatively low. Therefore, all brick walls, be they garden walls, interior room dividers or exterior veneer, must be solidly supported by a solid masonry foundation or footing. If

Fig. 206. *A beautiful example of a pierced brick wall made of concrete brick.* Courtesy National Concrete Masonry Association.

Fig. 207. *A single-brick wall given a serpentine shape for rigidity and beauty.* Courtesy National Brick Institute.

they are not, chances are they will crack sooner or later.

Outdoors. Brick footings must reach below the frost line. At a minimum, the footing should be 8 inches thick and at least twice as wide as the thickness of the wall it carries. Width can be reduced somewhat when the wall is low, but in the case of garden walls which are not normally braced against tipping, you have to make the footing wide enough to eliminate this possibility.

Exterior veneer can be laid up on an existing footing, adjoining an existing block wall. Or 4- or 6-inch concrete block can be laid up atop the footing, adjoining the existing block wall. The brick can then be laid up atop the added block. This is done to save on the cost of the brick. The block support is installed below grade, out of sight; just the brick veneer is visible.

Indoors. Brick can be laid directly on a concrete floor. All that need be done is to clean the floor of paint and oils. Brick cannot be laid atop a wood floor without the risk of the brickwork cracking. Wood moves with changes in season, changes in air moisture and the movement of people.

ESTIMATING BRICK QUANTITY

The accompanying table should provide all the data you require. Note that the joint widths greatly affect the total number of bricks you will require for a particular job. Note, too, that the estimated quantities of mortar for a given number of bricks include 10% additional mortar for waste. For a new mason, this figure is usually a bit low.

APPROXIMATE BRICK AND MORTAR QUANTITIES
NEEDED FOR WALLS OF VARIOUS DIMENSIONS

Wall area sq. ft.	Wall thickness					
	1 brick		2 bricks		3 bricks	
	Number of bricks	Mortar cu. ft.	Number of bricks	Mortar cu. ft.	Number of bricks	Mortar cu. ft.
1	6	.08	13	.2	19	.32
10	62	.8	124	2	185	3.2
100	617	8	1,233	20	1,849	32
200	1,234	16	2,466	40	3,698	64
300	1,851	24	3,699	60	5,547	96
400	2,468	32	4,932	80	7,396	128
500	3,085	40	6,165	100	9,245	160
600	3,712	48	7,398	120	11,094	192
700	4,319	56	8,631	140	12,943	224
800	4,936	64	9,864	160	14,792	256
900	5,553	72	10,970	180	16,641	288
1000	6,170	80	12,330	200	18,490	320

Brick and mortar quantities required for walls made of standard brick (2¼ by 3⅝ by 8 inches) using ½-inch-thick joints. When using ⅝-inch joints, multiply quantity by 80%. When using ⅜-inch joints, multiply quantity by 120%. Note, mortar figures include approximately 10% additional mortar for waste.

Fig. 208.

Mixing mortar, cutting and laying brick, etc., are all covered in Chapter 3.

GENERAL SUGGESTIONS

While a strong brick wall is easy to lay up, an attractive brick wall requires a little more care and patience. For the wall to be attractive all the joints must be of the same or nearly the same width; all the joints must be horizontal and vertical. This may sound difficult, but it is not if you take your time and do not say "It's good enough" and go on. Expect to remove and replace bricks a few times at the beginning (wash the bricks clean before reusing). With time you will gain the "eye" (as we masons call it) and you will be laying brick like an expert.

To help you keep joint width from varying as you go, keep a piece of wood of the same thickness as the desired joint width on hand. Every now and again, match the stick's width against your joint. In this way you will quickly see whether or not you are varying joint width. On a long wall, it is easy for a beginner to slowly change joint width as much as ¼ inch without being aware of the change.

Work with a ½-inch-wide joint rather than a ⅜-inch joint. More mortar is required, but variations in the wider width are less obvious.

And recess the joint. This is done with a joint-raking tool that removes the top half inch of mortar from the joint. In this way the joint is not directly visible from all positions, and joint variations are not readily noticeable. Rent or purchase the joint raker, or make one by driving a nail partway through a piece of wood. The nail's head is dragged down the length of the joint, recessing the joint.

Use a story pole. This is a length of wood positioned perfectly vertical near either end of the wall. The height of each succeeding course of brick is marked on the pole (Fig. 209). Thus, should you want to know where your seventh course falls in relation to where it should fall mathematically, check the course against the mark on the story pole. For example, laying 2¼-inch brick on a bed of ½-inch joints would put your tenth course at exactly 2 feet 3½ inches. Course twenty would reach to 4 feet 7 inches. If your horizontal joint width varied plus or minus, you would quickly see the change on the story pole. You can make your own pole with a rule and marking pen, or you can purchase a story pole premarked. You can also purchase a roll of pressure-sensitive tape premarked with course heights. The tape is then affixed to your pole.

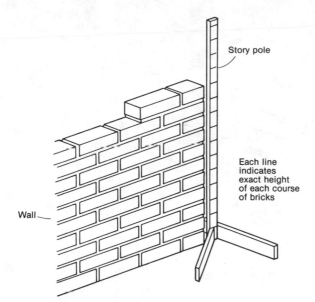

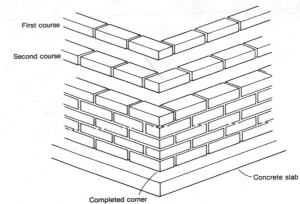

First course

Second course

Concrete slab

Completed corner

Fig. 210. A single-brick wall on a concrete slab.

Story pole

Each line indicates exact height of each course of bricks

Wall

Fig. 209. A story pole set up alongside a wall under construction. Marks on pole indicate exact desired heights of brick courses.

Keep your mortar on the dry side. Obviously, you won't secure proper adhesion if the mortar or the bricks are too dry, but if the mortar is too wet the bricks may sag and the chance of your smudging the face of the bricks as you work is greatly increased.

SINGLE-BRICK WALLS

Walls made of bricks laid end to end on their backs are called single-brick walls. They are used for garden walls, room dividers, planters, doghouses and even small cottages when constructed of 5½-inch-wide, SCR bricks.

Indoors, the brick can be laid up on a concrete floor; outdoors, the single-brick wall can be laid up on a concrete slab—e.g., the edge of a driveway or walk, assuming the slab is not subject to frost heave. If there is no slab, the single-brick wall needs an 8-inch-thick footing which rests on firm subsoil below the frost line, as usual. The width of the footing will depend on wall height. Wall height should be kept to a maximum of 4 feet above the footing when it is above grade (as in the case of a slab), and 4 feet above grade when the footing is below grade. Single-brick wall height can be increased when either SCR bricks are used or the wall is braced by one means or another.

RECOMMENDED MINIMUM FOOTING WIDTHS FOR SINGLE-BRICK, STRAIGHT WALLS

Footing width	Height above footing or grade
8 inches	2 feet
10	3
12	4

All footings to be a minimum of 8 inches thick.

Construction. Once the footing has been poured, or if you are working on an existing slab, simply lay up the bricks as suggested in Chapter 3. If the wall is a single straight line, start with the ends and work toward the middle. If the wall has turns or angles, treat the turn or angle as a corner and work from one end to the corner. Cap the brick, if you wish, with a layer of mortar either rounded or angled to shed water. Cap the pilasters and posts with slate or flagstone set in mortar at a slight pitch.

Pierced walls. Fig. 211 shows how bricks may be left out of a single-brick wall to produce various patterns and also permit the passage of air. When the opening is large, it is advisable to support the bricks

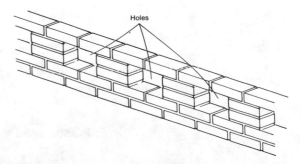

Holes

Fig. 211. Another example of a pierced, single-brick wall.

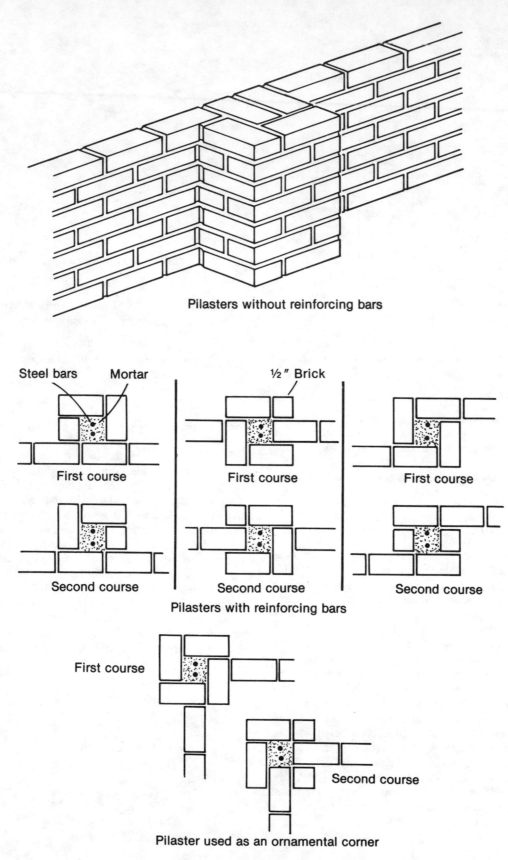

Pilasters without reinforcing bars

Steel bars   Mortar   ½″ Brick

First course   First course   First course

Second course   Second course   Second course

Pilasters with reinforcing bars

First course

Second course

Pilaster used as an ornamental corner

Fig. 212. Typical pilasters.

that make the span with some temporary support—blocks of wood, cardboard boxes or the like.

Adding pilasters. When a section of a wall is thickened to support a girder or to brace that wall, that section is called a pilaster (Fig. 212). It looks like a column embedded in the wall. Generally, the pilaster is constructed integral to the wall, meaning the bricks forming the pilaster overlap the bricks forming the wall. Sometimes pilaster bricks are closely spaced, leaving little or no vertical opening between them. Sometimes a space is left and the opening is filled with mortar. Sometimes the opening is filled with mortar plus a number of steel bars. The lower end of these bars may be embedded within the footing.

In any case, whatever the design details of the pilaster, it always permits the wall to be safely constructed to a greater height than is possible without the bracing of the pilaster. As can be seen in Fig. 212, a great number of factors are involved in the design, placement and spacing of pilasters that may be incorporated into a wall—so many that it is impractical to evaluate them here. Therefore, if you are going to use pilasters to safely increase the height of your wall, secure local engineering advice before doing so.

Angled walls. Another way to brace a single-brick wall is to introduce a change of direction every few feet. In other words, the wall and its footing weave back and forth in a series of equal, angular turns—zigzag. Again, there are too many possible variables to provide any simple rule of thumb to the safe height increase zigzagging provides, so it is a good idea to get local engineering advice first.

Planters. A planter is simply a four-wall, single-brick-thick structure. When constructed outdoors on earth, the footing is made square or rectangular to follow the shape of the box. Its center is left open. An on-soil, outdoor planter has no bottom, which permits it to drain. Built on a concrete slab, drain holes should be provided. This can be done by placing pieces of oiled ropes between a number of first-course bricks. When the mortar has set up, the ropes are removed, leaving holes (Fig. 213).

Construct the corners first. Check them for squareness and levelness the same way suggested for checking a building foundation. Then fill in the balance of the walls.

TWO-BRICK WALLS

Two-brick walls can be laid up any of three ways: solid, rowlock or spaced and insulated.

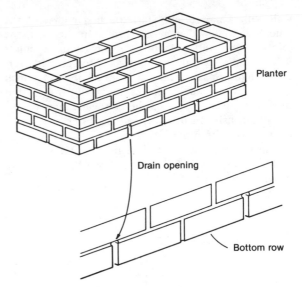

Fig. 213. Brick planter. If planter is constructed on a slab, drain holes (as shown) must be provided.

Footings. Footing requirements are essentially the same as for single-brick walls. The footing must rest on firm subsoil below the frost line. Since the two-brick wall is thicker than the single-brick wall, its footing must also be wider. Make the two-brick footing 8 inches wider than wall thickness, thus leaving a 4-inch shelf on each side of the wall. Make the footing a minimum of 8 inches thick when it supports no load, and half as thick as its width when it carries a load—as, for example, the roof of a building.

The two-brick wall can be laid up most easily and quickly as two individual, single-brick walls positioned back to back. Start the first course of one

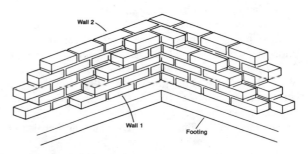

Fig. 214. A two-brick wall constructed of two single-brick walls placed back to back. Note that vertical joints are not in line.

wall with a half brick. Start the first course of the adjoining wall with a whole brick. In this way, the joints in one wall will not be in line with the joints in the adjoining wall, which makes for greater over-all strength. Space the walls from ¼ inch to 1 inch apart. A wider space makes the wall stronger but uses more mortar. Lay up the two walls more or less simultaneously. In this way, all the bricks will be in contact with fresh mortar.

Cap the two-brick wall with header bricks—bricks laid across the wall (Fig. 215). Pitch these bricks to provide water runoff.

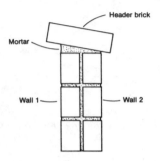

Fig. 215. One way a two-brick wall may be capped.

A stronger wall results when you bond the two walls together with headers. A bonded wall requires more care in construction, but it is much stronger, and some consider it more attractive.

Begin by deciding on what type of bond you are going to use. There are several. The difference between them is mainly appearance. The more common bonds are illustrated in Fig. 216. Next, start the first course of the wall with two whole bricks, laid up side by side and spaced just far enough apart to make the wall's width or thickness just equal to the length of the bricks being used. Complete the first course laying all the bricks in the stretcher position—lengthwise with the wall. Now, depending on the type of bond you have selected, lay the second and following courses. If you have chosen the common bond, the second course of bricks will all be laid up as headers, transverse to the length of the wall (Fig. 217). The following courses may all be stretchers up to the fourth, fifth or sixth course, which is again a course of headers.

Rowlock walls. This is a two-brick wall consisting of two spaced walls joined by a header every now and again, with at least one single-brick wall laid up with its bricks on edge as in Fig. 218. The result is an air space between the walls which provides a

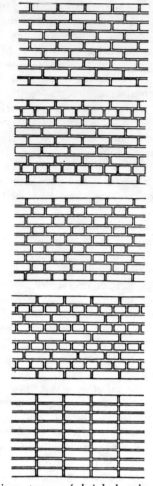

Fig. 216. Various types of brick bonds commonly used: A. Running Bond, B. Common Bond, C. Flemish Bond, D. English Bond, E. Stack Bond.

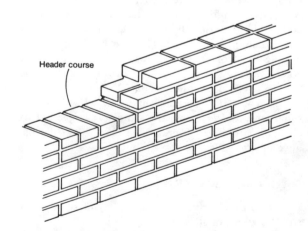

Fig. 217. An example of a two-brick bonded wall. The headers lying across the thickness of the wall increase its strength.

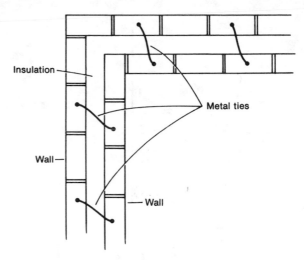

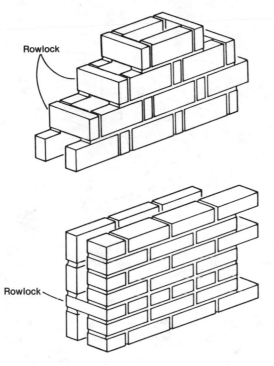

Fig. 218. Two types of rowlock walls.

Fig. 219. *An insulated wall. Space between bricks may be filled with insulation.*

little insulation, plus a stability produced by the thickness of the wall, with less brick than would be required for a solid two-brick wall. The rowlock bricks are positioned on edge, usually one rowlock header between every stretcher brick in each wall.

Insulated walls. These are two-brick walls constructed as two independent single-brick walls joined every 2 feet or so and every two courses by special metal ties (Fig. 219). The two walls are constructed simultaneously. The ties are simply laid in the mortar and covered with a brick. The space between the two walls may be filled with loose insulation. Since the thickness of the insulated two-brick wall is greater than the length of the bricks used, the bricks cannot be used to cap the wall. Instead, the wall may be capped with slate, flagstone or tile as previously suggested.

BRICK VENEER

Brick veneer is a single-brick wall constructed in front of any type of wall. The purpose of the brick veneer may be to protect the wall against the weather, as for example an exterior house wall of wood, or it may be to enhance the appearance of a wall. For example, a concrete-block basement wall

may be "hidden" with a veneer of brick to make a basement playroom more attractive. In any case, whatever the purpose of the veneer, the process of laying the brick is the same in almost all cases; only a few of the details may differ.

Veneering basement walls. Clean the wall and the adjoining 6 inches of concrete floor of all dirt, paint and lumps of old mortar. With the aid of a spirit level and a long straight-edged board, check the wall for verticality. If it is perfectly vertical, fine. It if tilts away from you, or toward you, you have to vary the position of the first course of bricks accordingly.

Start by applying Weld-Crete, Permaweld-Z or any other masonry bonding agent to the wall and the first 6 inches adjoining the bottom of the wall. If the wall is vertical, lay your first course of brick about ⅜ inch clear of the wall. If the wall tilts away from you, reduce this clearance as much as you can. If the wall tilts toward you, drop a plumb bob from the top of the wall to measure the tilt, then position the first course so that the top course of bricks will clear the top of the wall by ⅜ inch. In other words, begin the first course clear of the wall bottom by ⅜ inch plus the tilt distance (Fig. 220).

As you work, fill the space behind the bricks with mortar. Just let the mortar slide off your trowel. Do not force the mortar behind the bricks or you will push them out of line. Hide the joint at the top of the veneer and the ceiling with wood molding.

New building external veneer. The proper and easy installation of veneer on a new house begins with the footing. On the side or sides of the house

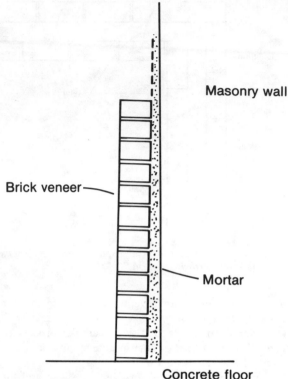

Masonry wall

Brick veneer —

Mortar

Concrete floor

Fig. 220. How a brick veneer wall must be positioned when the concrete or block wall it is to cover tilts outward.

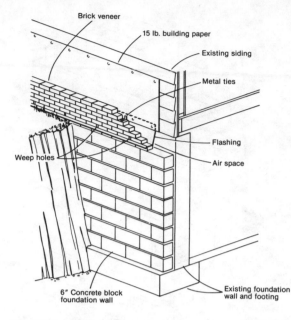

Brick veneer

15 lb. building paper

Existing siding

Metal ties

Flashing

Air space

Weep holes

6" Concrete block foundation wall

Existing foundation wall and footing

Fig. 221. How brick veneer may be supported by concrete block placed alongside a new or existing foundation wall.

Fig. 222. Side view of brick veneer resting on over-size foundation block.

that will be veneered, the footing is constructed 6 inches wider than elsewhere. The foundation blocks are laid up exactly as they normally would be. This done, an adjoining wall of 6-inch blocks is laid on top of the extra width of footing and bonded to the foundation wall with mortar (Fig. 221). The top of the 6-inch wall is terminated roughly 6 inches below the desired final grade. The brick veneer will rest on this 6-inch shoulder. As the shoulder stops before the surface of the earth, the joint between the bricks and the block will be out of sight when the house is completed and landscaped. Naturally, the 6-inch wall will be as long or a bit longer than the length of the veneer wall.

As an alternative, the foundation wall can be constructed of block 6 inches wider than required. Assume you need 8-inch block for the foundation. If you lay down 14-inch block up to the final course and use 8-inch block for that course, you will end up with a 6-inch-wide shelf on which you can rest your brick veneer wall.

Either way, the house is first framed and covered

with building paper, as it normally would be. Next, a 1-foot-wide strip of metal flashing is positioned against the building wall. The lower edge of the flashing rests on the block shelf. The top edge of the flashing is nailed to the building. One nail every 2 feet is plenty (Fig. 221).

Next, the brick is laid atop the 6-inch-block shoul-

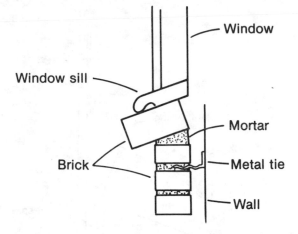

Fig. 223. One way brick veneer may be terminated beneath a windowsill. Space is filled with tilted brick set in mortar.

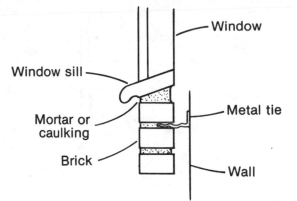

Fig. 224. When the top course of veneer works out correctly, you can fill the space with mortar or caulking.

der. The first course is laid down as you normally would, leaving about ½ inch of clearance between the rear of the brick and the flashing. In addition, a short piece of oiled rope is positioned between every fourth brick. Later, when the mortar has set up somewhat, the ropes are removed, leaving holes. These are called weep holes and they drain whatever moisture collects in the air space behind the brick. Succeeding courses are laid up normally; however, every so often the bricks are locked to the wall by means of a metal tie. The tie is a short corrugated piece of metal. One end of the tie is bent and nailed to the wall. The other end is positioned atop a brick and covered with mortar. The succeeding course of brick is laid atop the mortar. The tie does not interfere with the joint. Use one tie every 32 linear inches of wall and every 16 vertical inches, spaced apart so that no one tie is directly above another.

Under-window termination. If a portion of the top of the veneer wall works out so that a full brick can be fitted beneath a windowsill, you can simply seal the joint between the brick and wood with caulking. If the space is too large for one brick and too small for two, you can either cut the topmost course of bricks lengthwise, or you can angle the topmost course (Figs. 223 and 224).

Under-eave termination. The space between the topmost course of bricks and the soffit or the eave can be hidden with a piece of wood molding (Fig. 225).

Veneer-to-frame joints. When laying veneer up against a door- or window frame, always leave a fraction of an inch of clearance between the wood

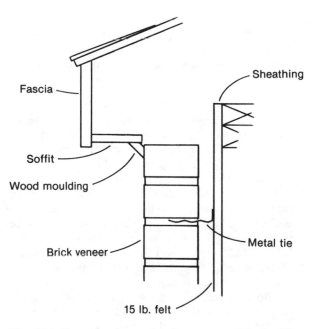

Fig. 225. Space between top of veneer and underside of an eave can be closed with wood molding.

and the brick to allow for movement. Then seal the opening with caulking.

Half-wall veneering. If, for the sake of appearance, you do not want the brick veneer to go higher than halfway up the side of an exterior wall, you must seal the opening between the veneer and the house siding against the weather. A simple, neat way of doing this consists of simply stopping when you have laid up the brick veneer as high as you wish;

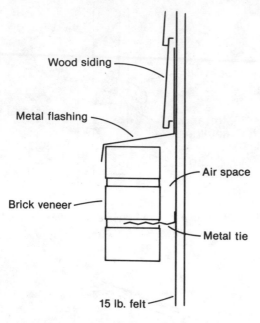

Fig. 226. *One way a half wall of veneer can be sealed against the weather.*

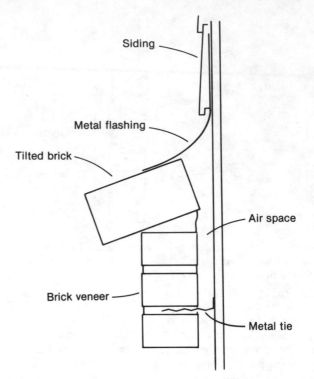

Fig. 227. *Another method that may be used to seal the top of a half wall of veneer against the weather.*

then an 8-inch-wide strip of metal flashing is folded lengthwise and placed atop the brick and against the house sheathing. A few nails hold it in place. Then the house siding is fastened in place on top of the flashing. No caulking is required.

Another, possibly more attractive method consists of stopping the brick veneer one course below the desired final height, then placing full-length or three-quarter-length brick at an angle across the in-place brick. The result is sort of a little roof. Again, a length of metal flashing is folded lengthwise, and the angle of the fold is adjusted to match the angle of the brick surfaces and the vertical wall. The flashing is placed atop the bricks and against the house sheathing, and the siding goes in place atop the flashing. The result is a more formal finish to the top of the veneer, but requires a lot more work and care.

Old buildings, external veneer. Excavate down to the footing alongside the side or sides of the building you wish to veneer with brick. Clean the footing and foundation wall of all soil. Let the asphalt waterproofing be; there is no practical way you can remove it. Give the bare masonry a coat of Permaweld-Z or any other masonry bonding agent. On the footing that extends beyond the foundation wall, lay up a wall of 6-inch block. As you go, fasten the new wall to the old with metal ties. Make the ties of ei-

ther ½-inch iron bars or 1 × ¼-inch pieces of strap iron bent into a U-shape (Fig. 228). Install one tie every 4 linear feet and every second course. Lock the bars in place by inserting them in holes cut in the old wall and the core holes in the new block, and then filling the holes with mortar. When your wall is completed, you can lay up the veneer on top. From the veneer-supporting wall on up, the procedure is exactly the same as described earlier.

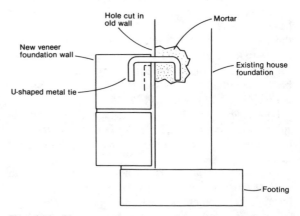

Fig. 228. *How metal ties may be used to lock a new veneer-supporting block wall to an old existing block wall.*

CHAPTER SEVENTEEN

Stoops and Porches

A stoop is an entrance platform outside the door of a building. Usually there are a number of steps leading down from the platform to a walk or driveway. Put a roof over the same platform and it becomes a porch. As far as we masons are concerned, the difference between porches and stoops is only one of size; porches are usually larger. But many municipalities do not take the difference as lightly as we do. Generally, municipalities do not consider a stoop to be an extension of the building, whereas a

Fig. 229. Masonry porch before final grading. Slab has been flagged with flagstone. Concrete block foundation wall has been parged. Rough texture was produced by holding the float loosely against the mortar while it was slowly rotated and pulled away.

porch is. This means that if your home is close to a side-yard line or a setback line, you may be able to construct a stoop that goes over the line without violating the local building code. However, if you roof over that entrance platform, there is a better than good chance you will be in difficulties with the local building department.

To be certain that neither your stoop nor porch will bring down the law, check with your building department before you add anything that will cross any of the building lines. Since the difference in masonry work between porches and stoops is negligible, we will henceforth refer to both as stoops.

BASIC PROBLEM

As explained in Chapter 11, Footings and Foundations, in order to construct both it is necessary to excavate a hole considerably larger than the final structure itself. When the footings and foundation have been erected, the space between the masonry and the old in-place virgin soil is filled with loose soil. This replaced soil is called fill. Whereas virgin soil (except the layer of humus atop it) cannot be appreciably compacted, fill settles and compacts by itself. Whereas you could construct a giant building directly on virgin soil without it subsiding (an exaggeration, of course), lay a single concrete block or stone on fill and you will see it subside and sink a little with each passing month. Since a stoop is positioned directly over fill, we have a basic problem: how can we keep the stoop from sinking into the earth?

Solutions. There are a variety of possible solutions to the problem of supporting a stoop. Your choice will depend on the size of the stoop, whether the building is still under construction or completed and whether or not you are willing to risk a little subsidence in exchange for lots of work.

SMALL STOOP
ON OLD FILL

When the fill has been in place twenty or more years, you can assume that it has been fairly compacted by the rain and that it will sustain a moderate load without subsiding. If you build the stoop so that

of the concrete you will pour. If a portion of the stoop will touch the wood portion of the building, isolate the concrete you will pour from the wood with a sheet of metal flashing. Nail the metal to the wood. Make the top edge of the metal higher than the top edge of the form.

To save on concrete, fill the center of the form with large stones, bricks and/or block. Use no wood or soil here. Arrange the central mound of stones so that its top is 6 inches below the top of the form,

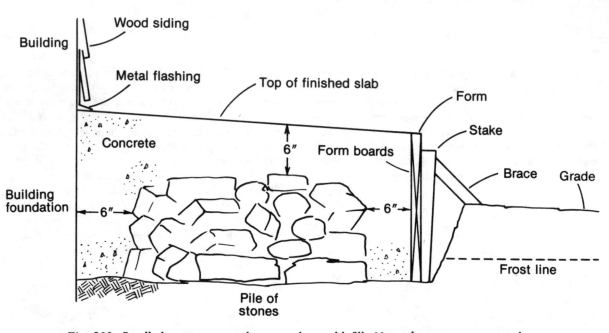

Fig. 230. Small, low stoop can be erected on old fill. Note that concrete goes down below frost line. Top of slab should be no more than 8 inches below doorsill.

it rests entirely on the earth rather than on a foundation and footings, load will be minimized and the chance of subsidence reduced. This design can be utilized with a stoop of any size, but it is only practical in terms of material when the frost line is close to the surface and the stoop is small.

Construction. Excavate a flat-bottomed hole a little larger than the desired stoop. Build a form of wood within the excavation. Let the building itself be one side of the form. Make the top of the form as high as the desired top of the stoop, and let the top of the form pitch away from the building at an angle of approximately ¼ inch to the foot. Take care to make the inside of the form smooth, as the inside surface of the form molds the external surface

and the sides of the mound are 6 inches or more clear of the sides of the form.

Pouring. Mix sufficient 1:2¼:3 concrete to fill the form to its top. Pour very slowly, taking care to fill the space between the central mound and the form wall evenly. Poke a stick or shovel into the concrete to aid it in settling into all the corners. After you have filled the space around the central mound of stones, wait twenty to thirty minutes to let this concrete set up a little. Then pour the balance of the form to its top. Puddle the concrete a bit with a shovel or stick to make it settle. Screed, tamp and float the usual way. Finish the edges with an edger. Wait at least two days in warm weather, a week in cold, before you remove the form. Wait a few weeks and then caulk the joint between the building and the stoop. The caulking goes over the metal flashing.

Provisions for installing railings and porch roof supports are covered at the end of this chapter. Steps are covered in the following chapter.

FOOTING-SUPPORTED STOOP

If your stoop is going to be much larger than 3 by 4 feet, you may find that it will be less costly in material to construct a footing-supported stoop. When you are not certain the soil next to your home has fully settled and stabilized, and you do not want to risk your handiwork cracking and moving away from the building, you have no choice but to build your stoop on a foundation that rests on a secure footing. Not only must this footing be below the frost line, it must also rest on virgin, undisturbed soil.

Foundation design. A porch stoop foundation is similar to the perimeter foundations and their foot- ings discussed in Chapter 11, but whereas a house foundation must of necessity always have four sides and form a complete enclosure, a stoop foundation does not. A stoop foundation need not have more than two sides. The basic difference between a two-sided stoop foundation and a four-sided foundation is in the slab they support. Slabs supported only on two sides need more steel and concrete than same-size slabs resting on four supports. Study the accompanying table and decide what design is best for you. Slabs smaller than those listed require no steel.

Stoop foundations, new construction. When you excavate for the building's footings, also excavate an adjoining area to accommodate the footing for the stoop. Decide whether you are going to support the stoop slab on four sides or less. If you are going to support the stoop slab where it meets the side of the house foundation wall, an adjoining, second foundation wall must be erected there, and it must, of course, rest on a footing. The best and easiest way to make this footing is to simply widen the house foot-

STEEL REINFORCEMENT FOR SLABS
SUPPORTED ONLY ON TWO SIDES

Slab Thickness	Span	Bar	Bar Spacing	Cross Bar	Cross Bar Spacing
4 inches	4 feet	⅜ inch	10 inches	⅜ inch	18 inches
4	5	⅜	8	⅜	18
4	6	⅜	6	⅜	18
4	8	½	7	⅜	12
4	10	½	4	⅜	6
6	4	⅜	12	⅜	18
6	5	⅜	10	⅜	18
6	6	½	10	⅜	14
6	8	½	7	⅜	12
6	10	½	5	⅜	10

STEEL REINFORCEMENT FOR SLABS
SUPPORTED ON FOUR SIDES

Slab Thickness	Span	Bar	Bar Spacing	Cross Bar	Cross Bar Spacing
4	5	⅜	18	⅜	18
4	6	⅜	14	⅜	14
4	8	⅜	10	⅜	10
4	10	½	10	½	10
6	5	⅜	20	⅜	20
6	6	⅜	18	⅜	18
6	8	⅜	14	⅜	14
6	10	½	12	½	12

NOTE: data is applicable only when deformed steel bars made for use with concrete are used, and load is residential.

(123)

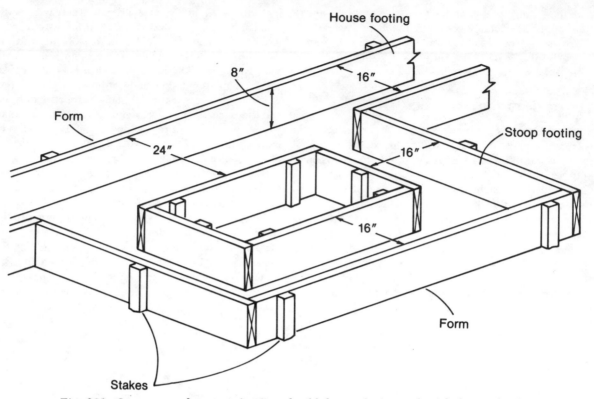

Fig. 231. On new work, stoop footing should be made integral with house footing. Note that house footing width alongside stoop has been made wide enough to support stoop foundation as well as house foundation.

ing to carry the stoop foundation wall. House footing width increase should equal the thickness of the block to be used for the stoop foundation wall. If you are not going to support the house side of the stoop slab, you need not erect a second foundation wall on the house footing.

To provide footing for the other three sides of the stoop foundation, construct a suitable footing form integral with the house footing. If possible, pour all the footing forms at one time.

If you are planning to drain the house foundation, cut holes through the sides of the stoop-footing form. Slip the drainpipe into place (holes in its sides), then cover the drainpipe in the form with tar paper so that concrete will not get into the pipe. Finally, pour the footing over it.

Stoop foundation, existing construction. To construct a stable stoop, you have no choice but to dig down to the level of the bottom of the existing footing. There you must construct the footing form. Butt its ends up against the old footing and make the top of the new footing flush with the old. If there is a

drainpipe alongside the old footing, cover it with tar paper and pour the footing over it.

If you are not going to support the house side of the slab, you do not need to erect a foundation wall adjoining the existing building foundation wall. If you are going to support the house side of the stoop slab, you will, of course, require a foundation wall beneath this side or edge of the stoop slab. This foundation wall (let's call it the fourth wall) is constructed on top of the building footing that extends beyond the house foundation wall. In most instances, this "shelf" is not going to be much more than 4 inches wide. Therefore, it is very important that the block you lay atop this shelf be firmly tied and bonded to the existing building foundation wall. This is done in two ways. First, the old wall is scrubbed clean of dirt and soil (there is nothing you can easily do about a layer of asphalt, so ignore it). Second, the bare block is given a coat of Permaweld-Z or a similar bonding agent. Third, metal ties are used to lock the new blocks to the old (Fig. 235). Make the ties from either ½-inch steel bars or 1 × ¼-inch

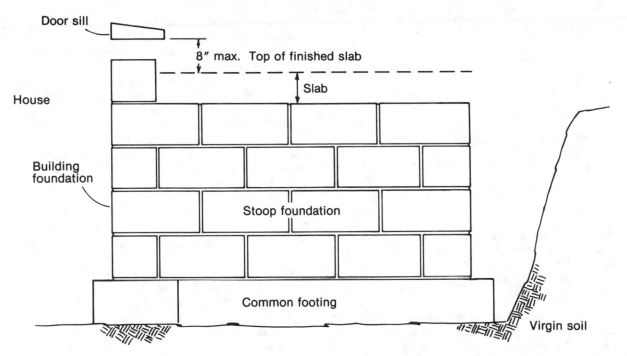

Fig. 232. Side view of new stoop construction.

Fig. 233. Stoop foundation almost completed. Mason is placing last block in position.

Fig. 234. Two-sided stoop foundation. Note provision for steps at ends of foundation.

(125)

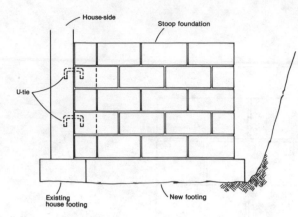

Fig. 235. Side view of stoop foundation added to existing foundation. New footing is simply butted up against existing footing.

strap iron bent into a U-shape. One tie every 4 linear feet and every 2 vertical feet of wall will be plenty. Fasten the ties by cutting holes with a chisel in the old wall, slipping the ties into the old wall and the new block, and locking the ties in place with mortar.

Making the stoop slab. Slab construction is the same whether the slab is supported on two, three or four sides; only the amount of reinforcement differs (see previous table).

With the stoop foundation wall erected, there is a hollow formed between them. To simply fill this space with concrete would be a waste of effort and money. The first step, therefore, is to provide a support, flush with the tops of the foundation walls, which will support the wet concrete. Do this by filling the space with rocks and soil, or by constructing a platform of wood supported by on-end concrete blocks. The space between the wood and the block may be sealed with mortar. The platform and its supports remain hidden by the slab forever.

Supporting-side computation. At this point you have erected the stoop foundation and are preparing to construct the form in which you will pour the stoop slab. The question may arise as to how many stoop-slab supporting sides there are. Obviously, two sides and four sides are self-evident. But what about three sides? Shall they be considered equal to two-sided support, or four? Since the slab table incorporates a generous safety factor, the following rules of thumb may be used.

If the slab is supported on four sides and no more

than 40% of any wall is open, it may be considered a four-sided support (a foundation wall is incomplete when an opening is left in its side to accommodate steps or a door leading to a storage area).

If two of the three supporting sides are each two times longer than the third, which is no more than 6 feet long, the slab may be considered to be supported on four sides.

If there are only two supporting sides and each is three times longer than the unsupported side, which is less than 6 feet long, calculate the slab as supported on four sides.

When planning a slab larger than 10 by 15 feet or so, even a small change in steel requirements amounts to a considerable sum. In such cases it is wise to secure engineering guidance.

Constructing the slab form. Use boards 2 or more inches wider than the desired thickness of the slab. Position the boards against the sides of the foundation, and hold them firmly in position with stakes and braces. Position the top edges of the form just as high above the top of the foundation as the required slab thickness. Pitch the form away from the building at an angle of ¼ inch to the foot. If there is to be steel within the slab, position the steel an inch or two closer to the bottom of the slab than its top. Raise the steel on stones. Tie the bars with wire where they cross over one another. Separate the concrete to be poured from the wood portion of the building with flashing. A few nails will hold it permanently in place.

If you want the slab to overhang the sides of the foundation, use spaces between the form sides and the foundation sides, as illustrated (Fig. 237).

Pouring and finishing. Use a high-sand mix, as it will be easier to finish. Pour slowly and carefully so as not to disturb the steel nor burst the form. Screed, tamp, edge and float-finish in the usual way. If you plan to flag the slab, just screed and tamp. The stoop slab may be flagged with slate or flagstone as described in Chapter 13.

Rail and roof posts. Wrought-iron railing posts are positioned with their ends within the slab. This can be accomplished in one of two ways. A hole can be provided in the concrete, or a hole can later be drilled in the concrete. The first method is simple enough but it requires care. A 2-inch-thick dowel is positioned within the wet slab, 4 inches in from the side or sides of the slab. The slab is finished. Then,

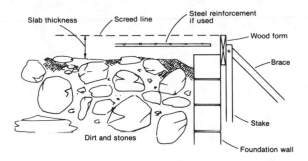

Fig. 236. *Cutaway view of stoop foundation wall and form used for containing the concrete that will be poured to make the slab. Here, the space between the foundation walls has been filled with rubble and dirt to the level of the top of the foundation walls.*

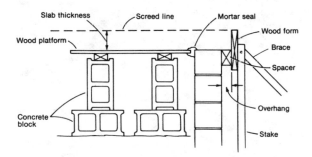

Fig. 237. *Same stoop foundation. Here, space between the walls has been closed off with a wood platform supported by block. Slab form is spaced away from foundation wall. This result is a slab that will overhang the foundation by the thickness of the spacer.*

Fig. 238. *Stoop form ready to be filled with concrete. Porch roof supporting column (top right) is of steel so it may be safely embedded in the slab.*

when the concrete has set up, but well before it has hardened, the dowel is carefully removed. If you drill, wait at least a month to make certain the concrete is hard. Then use a hollow-center, carbide-tipped, rotary drill. Keep the bit flooded with water, and use a drill just large enough to accommodate the post. Usually, these holes are made about 4 inches deep.

Nowadays, porch roof posts are not positioned with their ends within the concrete if the post is of wood. Instead, the bottom ends of the posts are fitted into metal cups having small pointed feet. Shallow holes are drilled into the concrete. The cup's points are placed within the holes. The weight of the porch roof resting on the post keeps it in place.

A hollow-center metal porch post can be rested in the correct location atop the concrete while it is still setting up. The weight of the post makes an indentation. The post is removed. Later, when the concrete has set up, the post is replaced within its indentation.

CHAPTER EIGHTEEN

Steps

Fig. 239. A beautiful example of a poured concrete staircase veneered with brick. Courtesy Structural Clay Products Institute.

Fig. 240. Measuring total rise of hill on which staircase will be constructed.

No matter what kind of steps you are planning to build—masonry, wood, etc.—your first move should always be to determine rise and tread. Rise is the height of one step above another; tread is the depth of the step. Rise should never be more than 8 inches. Higher makes for uncomfortable walking. Tread should never be less than 9 inches—less will prevent most people from placing their entire foot on the step. Outdoors, the preferable minimum is 12 inches.

DETERMINING RISE AND TREAD

Drive a peg into the top of the hill where the steps are to terminate. Drive a stake vertically into the bottom of the hill where the steps are to begin.

Stretch a horizontal line from the peg to the stake. Hang a line level in the center of the line to make certain it is truly horizontal (Fig. 240). The distance from the earth at the bottom of the stake to the line is the total rise. Divide this dimension into an equal number of parts—none more than 8 inches. Each part is, of course, the height of the riser.

If you are working with block, you will want each riser to be exactly 7⅝ inches because that is the exact height of a standard block. If riser dimensions do not work out to equal block, brick or any precut or cast masonry, think about making the first step

(128)

upward larger and deeper than the rest—a sort of platform. Since you can vary the height of this platform from 1 inch up to 8 inches in height, you can make the following riser heights match the dimensions of your masonry.

If you are going to construct a form and pour the steps of concrete, you can, of course, vary each riser to suit your purpose. However, if you can select a riser height that equals the width of a standard board, you will save yourself a lot of board cutting.

To find tread depth, measure along the line from peg to stake. Divide this figure by the number of steps you plan to make. As you can see, you are going to have to juggle the figures a bit until you find the best riser and tread dimensions for your staircase.

Excavation. If the hill is a gentle slope, your staircase can follow the contour of the hill. If the hill is steep, you have the choice of digging into the hillside to reduce the angle of the staircase, or you can build the stairs clear or partially clear of the hill. If the stairs lead to a building, terrace or stoop, the entire staircase will have to be exposed. This means you will have to take care that all the masonry work is neatly done. Buried in the side of a hill, you need only worry about the appearance of the steps themselves. Otherwise the construction remains exactly the same.

Fig. 241. Large front-walk steps made of flagstones laid on steps cut into the earth and stone risers.

DRY (MORTARLESS) STONE STEPS

You can construct steps from large, fairly flat fieldstones, flag- and bluestones and precast concrete slabs. Laid up without mortar, the steps will have both charm and durability. Without mortar, the construction is flexible and can rise and fall with the frost.

Dimensions. If you are working with flagstone and brick or block, you are working with fixed dimensions. Plan the risers and treads as suggested and vary them according to your needs by choice of brick, block and flagstone dimensions.

If you are working with fieldstones, it is catch-as-catch-can. The best you can do—without a lot of trouble—is to start your staircase at the bottom of the hill and work your way upward, letting the stones fall where they may. The risers will vary from step to step, and walking will therefore require a little care (we are accustomed to identical risers), but the results will be natural and most attractive.

Construction. The following procedure will be the same for whatever materials you may care to use. Only the dimensions will differ. Start at the bottom

of the hill. Cut into the hill to form a shelf that is flat, horizontal and pitched ¼ inch to the foot, front to back. At the same time, cut the shelf as deeply into the hillside as is necessary to secure the vertical dimension equal to the height of the following riser. Place one flat stone on the earth adjoining the vertical cut. Place whatever bricks or stones you need to produce the desired rise less the thickness of the tread stone that will follow on top of the flat stone which forms the first tread.

Cut a second shelf into the hillside. Make the bottom of this cut flush with the tops of the riser stones. Make the cut just as deep into the hill as the depth of the tread stone you will now lay down, and make the rear of the shelf a little higher than its front to provide the ¼-inch pitch for water runoff. Now place the second tread stone on top of this shelf. The front end of the stone overlaps and rests on the riser stones, locking them into place. Repeat this process until you reach the top of the hill.

SEMI-DRY STONE STEPS

Don't change a thing. However, when you position the riser stones, set them in mortar and cover their tops with mortar. The mortar helps hold the entire assembly together. And since you can easily vary the thickness of the layer of mortar you lay down, you can much more easily vary riser height. In addition, mortar makes it possible to use small, odd-shaped stones for the risers.

Side walls. If you have cut deeply into the hillside, you have exposed bare earth to either side of the staircase. If you wish, you can remove some of the earth and slope it gently away from the steps. Grass and ivy will hold the earth in place. Or you can build small stone walls alongside the steps, as described in Chapter 14. These stones can be laid up dry or with mortar.

MORTARED STEPS

Stone steps. These steps may be constructed the same way as semi-dry steps, the difference being that mortar is used between all the stones, including the stones forming the side walls. In the case of the latter, the side stones are usually placed atop the edge of the tread and merely leaned against the earth alongside. Obviously, this type of construction will crack when hit by frost. But since it is informal, a few cracks here and there usually do not mar its appearance. Since the stones remain in place by virtue of their weight, cracks do not materially weaken the structure.

Fig. 243. Here, masons are lowering flagstone tread onto a bed of mortar.

Fig. 244. Mason is laying up a cut stone riser on a bed of mortar. Note the block behind the stone.

Fig. 242. Partially cut fieldstones laid up in mortar directly on the earth. In some places stone chips and small stones were laid down to provide a base for the tread stones.

The alternative to free-form construction is laborious, time-consuming and beautiful. The steps and side walls are constructed of concrete block or poured concrete carried on a solid, below-frost-line foundation and footing. Then the surface of the concrete and/or block is veneered with cut stone. This makes for a permanent structure that can easily last centuries.

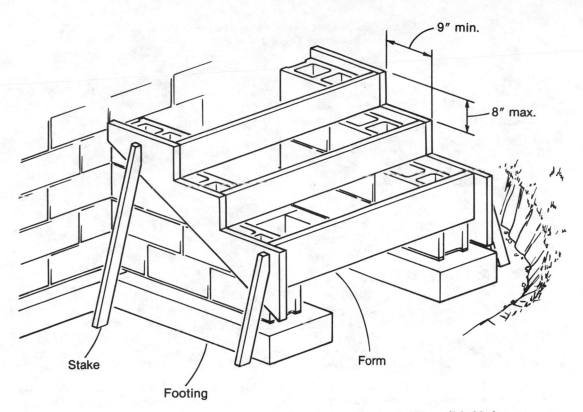

9" min.

8" max.

Stake

Footing

Form

Fig. 245. How concrete block steps are used to support the forms that will hold the concrete used to make the step treads. Space between block step walls must be filled with rocks and soil.

Block steps. The steps, actually a staircase, consist of two facing stepped walls of block. The top of each horizontal block serves to support one end of a tread (Fig. 245). This can be poured in place, or it may be a precase concrete tread, mortared into place. The two block step walls must rest on a solid concrete footing, which in turn must rest on virgin soil below the frost line. Normally, block staircase footings are poured when the house or stoop footings are poured. Usually the steps are constructed integral with the foundation or stoop wall. This simply means that the junction between the step wall and the other wall is treated like a corner; one block overlaps another.

Construction. With the footing in place, determine the desired elevation of the surface of the topmost tread. You want this surface to be 8 inches or less beneath the surface of an adjoining stoop or doorsill. Assuming you lay up the block with ⅜-inch-thick joints, each block will be exactly 8 inches high. Thus, the topmost tread on the staircase will be an even multiple, plus the 4 inches (usual thickness) of the last tread. Knowing this, you can measure up from the top of the footing and see where the top of

the last step will be. If this elevation is incorrect in relation to either a doorsill or what will be the finished top of a stoop, make the correction with the first course of block. If the topmost step works out to be too high, cut the first course of block lengthwise accordingly. If too low, raise the first course on mortar or bricks or similar masonry supports. If you have cut the first course, the first step up will have a lower riser. If you have raised the first course, the riser of the first step up will be higher than the others. Solve this by placing a thick slab of flagstone in front of the first step, or a slab of concrete made with the aid of a form.

Form. At this point you have two parallel stepped walls of concrete block with a large open space between. You need to fill this space with soil, loose rock or block to form a hill. The face of the hill is angled to match the general angle of the staircase. When the step form you are going to build and position is in place, the face of the hill should not be less than 6 inches from any form board.

Use 1-inch boards to construct the form illustrated. Press them tightly against the sides of the block and hold them there with stakes and braces.

Fig. 246. The block steps have been erected. The step form has been constructed and fastened in place. The slab form is also ready and in place. Now the masons are ready to pour the slab.

Fig. 248. The stoop slab has been poured. Mason is spreading concrete; screed is to his left. Mix is much too wet.

Fig. 247. Looking down between the sides of the stoop. The space has been filled with rubblestone.

Fig. 249. Masons are now filling step forms with concrete.

Fig. 250. Using floats to screed the concrete surfaces level with the tops of the step form.

Fig. 252. Parging the sides of the staircase. Rough texture is produced by pulling trowel away from mortar.

Fig. 251. Form has been removed. Note how stair treads project beyond the block supports.

Position the boards so as to produce 4-inch-thick treads, the tops of which pitch down and away from the top of the steps at an angle of about ¼ inch to the foot. Some masons also give the stairs a side pitch to help the water run off to the side of the staircase.

Reinforcement. When each step tread is more than 5 feet long or 16 inches wide, position two ⅜-inch steel bars within the form for each tread. More or less center the bars lengthwise within each tread.

When you are going to build more than six steps, it is advisable to provide vertical reinforcement. Position one ½-inch steel bar for every foot of staircase width several inches above the sloped hill. Cross these bars with one ⅜-inch bar every 2 feet or so. Use iron wire and bind the bars together. Use rocks to lift the assembly into position. When the staircase is poured, these bars will be within the concrete.

Pouring and finishing. Use a 1:2¼:3 mix a little on the stiff side. Mix a little at a time. Carefully place the concrete behind the lowest step form board. Then, with a small board, screed the concrete level with the form board. Tamp it gently. When the first step has partially set up, you can pour the second, and so on. If the mix is too wet it will run out of the lowest step. If you fill the form too quickly it will burst. Float and edge the steps as soon as they are ready.

With great care, remove the form a bit early. Use your float and a little mortar to touch up and fill the rough spots. Plaster the sides of the staircase with mortar, and the job is done.

ALL CONCRETE STEPS

In the previous example we constructed two stepped walls of block. The block did several things for us. It formed the sides of the staircase, and it positioned and supported each step tread. More important, the block supported the form. In other words, the form was relatively easy to construct.

Fig. 253. Small staircase made of block. It will now be given a final side-wall coat of cement plaster. Slab and treads will be flagged with flagstone.

Fig. 254. An all-concrete staircase form prior to pouring. Note the steel bars. These continue on up and over the slab area.

Fig. 255. An all-concrete staircase veneered with Belgian block and flagstone.

Fig. 256. A very gently sloping staircase made of flagstone laid atop a concrete base. Border consists of Belgian block also laid up on concrete.

When you construct an all-concrete staircase, you must construct the entire form, in other words, the blocks are replaced by wood. Since the height of the concrete within the form will be higher than in the other designs, concrete pressure on the form will be greater and the form must be stronger and filled even more slowly.

Construction. Begin by excavating and building a concrete footing, which, as stated many times previously, must rest on firm, virgin soil below the frost line. Build the form as illustrated. Do not spare the stakes and braces. Leave 6 inches of clearance between the central hill or mound of earth and stones in the center of the form and its sides and steps. Follow the suggestions for reinforcement given previously. You can use a 1:3:5 mix for the out-of-sight portion of the staircase, if you wish to save money. But for the visible portion, use the high-sand 1:2¼:3 mix. It will finish better and more easily. Pour very slowly and carefully. Screed, tamp, float and edge the steps. Remove the form a bit early, again with great care. Then, with the float and some fresh mortar, touch up the hollows and rough spots. If you have used smooth boards for the form, there should not be any need to plaster the sides of the staircase.